Student Solutions Ma

Calculus

with Analytic Geometry

Fourth Edition

C. H. Edwards, Jr.

David E. Penney

Prentice Hall, *Englewood Cliffs, New Jersey 07632*

Production Editor: *Kris Ann E. Cappelluti*
Production Coordinator: *Trudy Pisciotti*

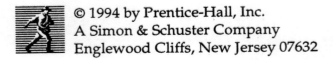

Printed in the United States of America

10 9 8 7 6 5 4 3 2 1

ISBN 0-13-457953-4

Prentice-Hall International (UK) Limited, *London*
Prentice-Hall of Australia Pty. Limited, *Sydney*
Prentice-Hall Canada Inc., *Toronto*
Prentice-Hall Hispanoamericana, S.A., *Mexico*
Prentice-Hall of India Private Limited, *New Delhi*
Prentice-Hall of Japan, Inc., *Tokyo*
Simon & Schuster Asia Pte. Ltd., *Singapore*
Editora Prentice-Hall do Brasil, Ltda., *Rio de Janeiro*

CONTENTS

PREFACE

This manual contains solutions to problems numbered 1, 4, 7, 10, ... in *Calculus with Analytic Geometry,* 4th edition (1994), by C. H. Edwards, Jr. and David E. Penney. Because answers to most odd-numbered problems are given in the *Answer* section of the text, corresponding solutions are omitted from this manual whenever the answer alone to such a problem should suffice. If the answer alone to an even-numbered problem is sufficient, we give here only the answer. When the solution consists of a routine verification of a formula, we generally omit it. When a suggestion should be enough for the solver, frequently that's all that appears. On a few occasions when the solution would spoil the problem for the solver, we omit it.

Many calculus problems can be solved by a variety of methods. We have tried to use the most natural method whenever possible, and only in rare instances have we put a clever method in its place, and then only when its educational value justifies such substitution. In particular, we used in working end-of-section problems only those techniques already developed in preceding material.

We gratefully acknowledge the encouragement, advice, and assistance of our colleagues and students in helping us to correct errors in earlier editions of this manual. In particular, we want to thank Terri Bittner of Laurel Tutoring (San Carlos, California), who with her staff checked every worked-out example and odd-numbered answer in the text, thus uncovering several errors in the earlier editions of this manual. We would also like to thank Deborah Sherman, who checked most of the solutions in the second edition of this manual. (New solutions were for the most part checked using *Derive* or *Mathematica*.) We would also like to thank Will Kazez and Ted Shifrin for invaluable assistance with TEX and *Adobe Illustrator,* the two programs used to typeset and create the figures for this manual. Most of all, we thank Carol W. Penney for typesetting the entire manual.

We took extra care in checking this manual for errors, but it seems inevitable that some will remain. You can call them to our attention by writing to either of us at the Department of Mathematics, University of Georgia, Athens, GA 30602-7403 or by e-mail.

C. H. Edwards, Jr.   (hedwards@math.uga.edu)
David E. Penney   (dpenney@math.uga.edu)

Typeset by AMS-TEX

Chapter 1: Functions and Graphs

Section 1.1

1. $|3 - 17| = |-14| = 14$

4. $|5| - |-7| = 5 - 7 = -2$

7. $|(-3)^3| = |-27| = 27$

10. $-|7 - 4| = -|3| = -3$

13. (a) $f(-a) = \dfrac{1}{a^2 + 5}$
    (b) $f(a^{-1}) = \dfrac{a^2}{1 + 5a^2}$;
    (c) $f(\sqrt{a}) = \dfrac{1}{a + 5}$;
    (d) $f(a^2) = \dfrac{1}{a^4 + 5}$

16. $g(a) = 5$: $\dfrac{1}{2a - 1} = 5$; $2a - 1 = \frac{1}{5}$; $2a = \frac{6}{5}$; $a = \frac{3}{5}$

19. $\sqrt[3]{a + 25} = 5$; $a + 25 = 125$; $a = 100$

22. $f(a + h) - f(a) = 1 - 2(a + h) - (1 - 2a) = -2h$

25. $f(a + h) - f(a) = \dfrac{1}{a + h} - \dfrac{1}{a} = \dfrac{a - (a + h)}{a(a + h)} = -\dfrac{h}{a(a + h)}$

28. $f(n/3) = n$ for every integer n, so f takes on all integral values. Because f can assume only integral values, its range is the set Z of all integers.

31. The set $\mathbb{R}$ of all real numbers

34. The set $[0, \infty)$ of all nonnegative real numbers, because $\sqrt{t}$ is defined only for $t \geq 0$.

37. $1 - 2t \geq 0$: $t \leq \frac{1}{2}$

40. We require $3 - t \geq 0$ and $3 - t \neq 0$, thus $t < 3$.

43. We require $x \geq 0$ and $4 - \sqrt{x} \geq 0$. The latter implies $x \leq 16$, so the domain of f is the interval $[0, 16]$.

46. If a square has perimeter P then each of its edges has length $P/4$, so the area of the square is:

$$A(P) = \tfrac{1}{16}P^2, \qquad P \geq 0 \quad (\text{or } P > 0).$$

49. $C(F) = \frac{5}{9}(F - 32), \quad F > -459.67$.

52. If x new wells are drilled, there will be $20 + x$ wells in all, each producing $200 - 5x$ barrels of oil per day. Hence the total daily production p of the oil field will be $p(x) = (20 + x)(200 - 5x)$, with domain consisting of all integers x in the range $0 \leq x \leq 40$.

55. If the cylinder has height h, then $\pi r^2 h = 1000$. The total surface area of the cylinder is $2\pi r^2 + 2\pi rh$, so

$$A(r) = 2\pi r^2 + \frac{2000}{r}, \qquad r > 0.$$

58. $A(x) = x(50 - x)$, $0 \leq x \leq 50$. Here is a table of a few values of the function A at some special numbers in its domain:

x	0	5	10	15	20	25	30	35	40	45	50
A	0	225	400	525	600	625	600	525	400	225	0

It appears that when $x = 25$ (so the rectangle is a square), the rectangle has maximum area 625.

61.

x	0.0	0.2	0.4	0.6	0.8	1.0
y	1.0	0.44	-0.04	-0.44	-0.76	-1.0

x	0.20	0.25	0.30	0.35	0.40
y	0.44	0.3125	0.19	0.0725	-0.04

x	0.35	0.36	0.37	0.38	0.39	0.40
y	0.0725	0.0496	0.0269	0.0044	-0.0179	-0.04

From this point on, the data for y will be rounded.

x	0.380	0.382	0.384	0.386	0.388	0.390
y	0.0044	-0.0001	-0.0045	-0.0090	-0.0135	-0.0179

Answer (rounded to two places): 0.38. The quadratic formula yields the two roots $\frac{1}{2}(3 \pm \sqrt{5})$; the smaller of these is approximately 0.381966011250105151795.

64. -3.24

67. 3.21

70. -9.28

Section 1.2

1. Both the segments AB and BC have the same slope 1.

4. AB has slope 1 but BC has slope $\frac{8}{7}$.

7. AB has slope 2 and AC has slope $-\frac{1}{2}$.

10. $m = -1$, $b = 1$

13. $m = -\frac{2}{5}$, $b = \frac{3}{5}$

16. The points $(2, 0)$ and $(0, -3)$ lie on L, so the slope of L is $\frac{3}{2}$. An equation of L is $2y = 3x - 6$.

19. The slope of L is -1; an equation is $y - 2 = 4 - x$.

22. The *other* line has slope $-\frac{1}{2}$, so L has slope 2 and therefore equation $y - 4 = 2(x + 2)$.

25. The parallel lines have slope 5, so the line $y = -\frac{1}{5}x + 1$ is perpendicular to them both. It meets the line $y = 5x + 1$ at $(0, 1)$ and the line $y = 5x + 9$ at $(\frac{-20}{13}, \frac{17}{13})$, and the distance between these two points is
$$\sqrt{\left(\tfrac{20}{13}\right)^2 + \left(\tfrac{4}{13}\right)^2} = \tfrac{4}{13}\sqrt{26} \approx 1.568929.$$

28. Show that AB is parallel to CD, that BC is parallel to AD, that AB and CD have the same length, that AD and BC have the same length, and that AC is perpendicular to BD by showing that the product of their slopes is -1.

31. The slope of $P_1 M$ is
$$\frac{y_1 - \bar{y}}{x_1 - \bar{x}} = \frac{y_1 - \frac{1}{2}(y_1 + y_2)}{x_1 - \frac{1}{2}(x_1 + x_2)} = \frac{2y_1 - y_1 - y_2}{2x_1 - x_1 - x_2} = \frac{y_1 - y_2}{x_1 - x_2},$$
and the slope of MP_2 is the same.

34. Begin with the fact that $L = mC + b$ for some constants m and b. The data given in Problem 34 yield
$$124.942 = 20m + b \text{ and } 125.134 = 110m + b.$$
It follows that $m = \frac{192}{90000}$ and $b = \frac{187349}{1500}$. Thus, approximately, $L = (0.00213)C + 124.899$.

37. $x = -2.75$, $y = 3.5$

40. $x = \frac{25}{4}$, $y = \frac{3}{2}$

43. $x = -\frac{7}{4}, \quad y = \frac{33}{8}$

46. $x = \frac{59}{11}, \quad y = -\frac{12}{11}$

Section 1.3

1. $x^2 - 4x + 4 + y^2 = 4$: $(x-2)^2 + (y-0)^2 = 2^2$. Center $(2,0)$, radius 2.

4. $x^2 + 10x + 25 + y^2 - 20y + 100 = 25$: $(x+5)^2 + (y-10)^2 = 5^2$. Center $(-5,10)$, radius 5.

7. $y = (x-3)^2$: Opens upward, vertex at $(3,0)$.

10. $2y = x^2 - 4x + 4 + 4$: $y - 2 = \frac{1}{2}(x-2)^2$. Opens upward, vertex at $(2,2)$.

13. $x^2 - 6x + 9 + y^2 + 8y + 16 = 0$: $(x-3)^2 + (y+4)^2 = 5^5$. Circle, center $(3,-4)$, radius 5.

16. $x^2 + y^2 - x + 3y + 2.5 = 0$: $x^2 - x + 0.25 + y^2 + 3y + 2.25 = 0$: $(x-0.5)^2 + (y+1.5)^2 = 0$. The graph consists of the single point $(0.5, -1.5)$.

19. The graph is shown below.

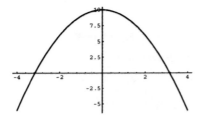

22. The graph is shown below.

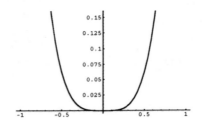

25. The domain of f consists of those numbers x such that $|x| \geq 3$. The graph is shown below.

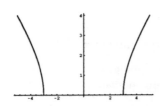

28. The graph is shown below.

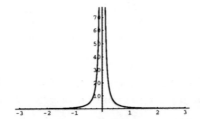

31. The graph is shown below.

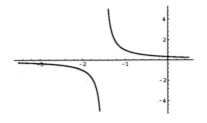

34. The graph is shown below.

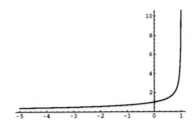

37. The graph is shown below.

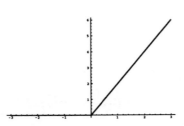

40. The graph is shown below.

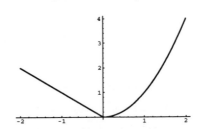

43. The graph is shown below.

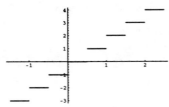

46. The graph is shown below.

49. $(2.25, 1.75)$

52. $(-3.4, 4.8)$

55. $y = -16(t^2 - 6t) = -16(t^2 - 6t + 9) + 144 = -16(t - 3)^2 + 144$. Answer: 144 feet.

58. $P = f(x) = 4000 + 100x - 5x^2 = 4500 - 5(x - 10)^2$, so $x = 10$ clearly maximizes P; $f(10) = 4500$.

Section 1.4

1. $(f + g)(x) = x^2 + 3x - 2$, $(f \cdot g)(x) = x^3 + 3x^2 - x - 3$, and $(f/g)(x) = \dfrac{x + 1}{x^2 + 2x - 3}$. The first two have domain the set $\mathbb{R}$ of all real numbers. The third has domain consisting of the set of all real numbers other than 1 and -3.

4. $(f + g)(x) = \sqrt{x + 1} + \sqrt{5 - x}$, $-1 \le x \le 5$; $(f \cdot g)(x) = \sqrt{5 + 4x - x^2}$, same domain.

    $(f/g)(x) = \sqrt{\dfrac{x + 1}{5 - x}}$, $-1 \le x < 5$

7. $(f + g)(x) = x + \sin x$ with domain $\mathbb{R}$; $(f \cdot g)(x) = x \sin x$, same domain;

    $(f/g)(x) = \dfrac{x}{\sin x}$; domain the set of all real numbers not integral multiples of π.

10. $(f + g)(x) = \cos x + \tan x$. The domain of $f + g$ consists of those real numbers x such that x is not an odd integral multiple of $\pi/2$. $(f \cdot g)(x) = \cos x \tan x$ with the same domain as $f + g$.

    $(f/g)(x) = \dfrac{\cos x}{\tan x}$. The domain of f/g consists of those real numbers x such that x is not an integral multiple of $\pi/2$.

13. Fig. 1.4.21

16. Fig. 1.4.22

19. Fig. 1.4.28

22. Fig. 1.4.26

25. 3

28. 2

31. 1

34. 1

1. $x \geq 4$

4. The set $I\!R$ of all real numbers

7. $x \leq \frac{2}{3}$

10. $[2, 4]$

13. First, $R = E/I = 100/I$, so $25 < 100/I < 50$; $\frac{1}{50} < I/100 < \frac{1}{25}$; $2 < I < 4$.

16. If the cylinder has radius and height x, then its volume is $V = \pi x^3$ and its total surface area is $S = 2\pi x^2 + 2\pi x^2 = 4\pi x^2$; but $x = (V/\pi)^{1/3}$, so $S = S(V) = 4\pi (V/\pi)^{2/3}$, $0 < V < \infty$.

19. $y - 5 = 2(x + 3)$

22. $2y = 3x - 12$

25. The graph is shown below.

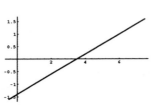

28. The graph is shown below.

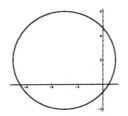

31. The graph is shown below.

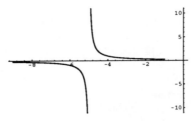

34. The graph is shown below.

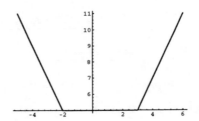

37. If $x - 3 > 0$ and $x + 2 > 0$, then $x > 3$ and $x > -2$, so $x > 3$. If $x - 3 < 0$ and $x + 2 < 0$, then $x < 3$ and $x < -2$, so $x < -2$. Answer: $(\infty, -2) \cup (3, \infty)$.

40. $2x \geq 15 - x^2$: $x^2 + 2x - 15 \geq 0$, so $(x - 3)(x + 5) \geq 0$. Now $x + 5 > x - 3$, so $x - 3 \geq 0$ or $x + 5 \leq 0$. Thus $x \geq 3$ or $x \leq -5$. Answer: $(-\infty, -5] \cup [3, \infty)$.

43. 1.191, 2.309

46. -9.962, 1.740

49. $(7/4, -5/4)$

52. $(-37/9, 35/9)$

55. Three solutions.

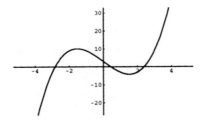

58. Two solutions.

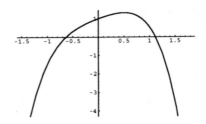

1. $f(x) = 0 \cdot x^2 + 0 \cdot x + 5$, so $f'(x) = 2 \cdot 0 \cdot x + 0 = 0$.

4. $f'(x) = -4x$

7. Take $a = 2$, $b = -3$, and $c = 4$. Then $f'(x) = 4x - 3$.

10. $f(x) = 15x - 3x^2$, so $f'(x) = 15 - 6x$.

13. $f(x) = 4x^2 + 1$, so $f'(x) = 8x$.

16. $dy/dx = 10 - 2x$; horizontal tangent at $(5, 25)$.

19. $y = x - \dfrac{x^2}{100} = x - \frac{1}{100}x^2$, so $dy/dx = 1 - \frac{1}{50}x$; thus there is a horizontal tangent at $(50, 25)$.

22. $$\frac{f(x+h) - f(x)}{h} = \frac{(x+h)^2 - (x+h) - 2 - (x^2 - x - 2)}{h}$$
    $$= \frac{x^2 + 2xh + h^2 - x - h - 2 - x^2 + x - 2}{h}$$
    $$= \frac{2xh + h^2 - h}{h} = 2x + h - 1 \to 2x - 1 \text{ as } h \to 0.$$

    Thus $f'(x) = 2x - 1$. The slope at $(2, f(2))$ is 3; the tangent there is $y = 3(x - 2)$.

25. $$f'(x) = \lim_{h \to 0} \frac{f(x+h) - f(x)}{h}$$
    $$= \lim_{h \to 0} \frac{(x + h - 1)^2 - (x - 1)^2}{h}$$
    $$= \lim_{h \to 0} \frac{x^2 + 2xh + h^2 - 2x - 2h + 1 - x^2 + 2x - 1}{h}$$
    $$= \lim_{h \to 0} \frac{2xh + h^2 - 2h}{h}$$
    $$= \lim_{h \to 0} (2x + h - 2) = 2x - 2.$$

    Now $f'(2) = 2$ and $f(2) = 1$, so an equation for the tangent line at $(2, f(2))$ is $y - 1 = 2(x - 2)$.

28. $$f'(x) = \lim_{h \to 0} \frac{(x + h)^4 - x^4}{h} = \lim_{h \to 0} \frac{x^4 + 4x^3h + 6x^2h^2 + 4xh^3 + h^4 - x^4}{h}$$
    $$= \lim_{h \to 0} (4x^3 + 6x^2h + 4xh^2 + h^3) = 4x^3.$$

    The tangent line at $(2, 16)$ has slope $f'(2) = 32$, and thus equation $y - 16 = 32(x - 2)$.

31. $$f'(x) = \lim_{h \to 0} \frac{\dfrac{2}{x + h - 1} - \dfrac{2}{x - 1}}{h} = \lim_{h \to 0} \frac{2(x - 1) - 2(x + h - 1)}{h(x + h - 1)(x - 1)}$$
    $$= \lim_{h \to 0} \frac{-2h}{h(x + h - 1)(x - 1)} = \lim_{h \to 0} \frac{-2}{(x + h - 1)(x - 1)} = -\frac{2}{(x - 1)^2}.$$

34. $dy/dx = -1 - 4x$. The slope at P is 3, so an equation for the tangent line there is $y - 4 = 3(x + 1)$, and an equation for the normal line there is $y - 4 = -\frac{1}{3}(x + 1)$.

37. $y'(t) = 96 - 32t$; $y'(t) = 0$ when $t = 3$. Therefore the maximum height attained by the ball is $y(3) = 144$ (ft).

40. The projectile hits the ground when $y = 0$, which occurs when $x - (x/25)^2 = 0$: $x = 0$ or $x = 625$. Therefore the projectile travels a horizontal distance of 625 feet. Its maximum height is attained when $dy/dx = 0$; that is, when $1 - \dfrac{2x}{625} = 0$: $x = 312.5$, and so $y_{\max} = 156.25$ (ft).

43. Suppose that (a, a^2) is the point on the graph of $y = x^2$ closest to $(3, 0)$. Let L be the line segment from $(3, 0)$ to (a, a^2). Under the plausible assumption that L is normal to the tangent line at (a, a^2),

we infer that the slope m of L is $-\dfrac{1}{2a}$ because the slope of the tangent line is $2a$. Because we can also compute m by using the two points known to lie on it, we find that

$$m = -\frac{1}{2a} = \frac{a^2 - 0}{a - 3}.$$

This leads to the equation $0 = 2a^3 + a - 3 = (a - 1)(2a^2 + 2a + 3)$, which has $a = 1$ as its only real solution. Intuitively, it's clear that there *is* a point on the graph nearest $(3, 0)$, so we have found it: That point is $(1, 1)$.

Alternatively, if (x, x^2) is an arbitrary point on the given parabola, then the distance from (x, x^2) to $(3, 0)$ is the square root of $f(x) = (x^2 - 0)^2 + (x - 3)^2 = x^4 + x^2 - 6x + 9$. A positive quantity is minimized when its square is minimized, so we minimize the distance from (x, x^2) to $(3, 0)$ by minimizing $f(x)$. But $f'(x) = 4x^3 + 2x - 6 = 2(x - 1)(2x^2 + 2x + 3)$, and (as before) the equation $f'(x) = 0$ has only one real solution, $x = 1$. Again appealing to intuition for the existence of a point on the parabola nearest to $(3, 0)$, we see that it can only be the point $(1, 1)$.

46. 3

49. -1

Section 2.2

1. $\displaystyle\lim_{x \to 0} (3x^2 + 7x - 12) = 3 \left(\lim_{x \to 0} x \right)^2 + 7 \left(\lim_{x \to 0} x \right) - \lim_{x \to 0} 12 = 3 \cdot 0^2 + 7 \cdot 0 - 12 = -12.$

4. $\displaystyle\lim_{x \to -2} (x^2 - 2)^5 = \left(\lim_{x \to -2} (x^2 - 2) \right)^5 = (4 - 2)^5 = 32$

7. $\displaystyle\lim_{x \to -1} \frac{x + 1}{x^2 - x - 2} = \lim_{x \to -1} \frac{x + 1}{(x + 1)(x - 2)} = \lim_{x \to -1} \frac{1}{x - 2} = -\frac{1}{3}$

10. $\displaystyle\lim_{y \to 3} \frac{\dfrac{1}{y} - \dfrac{1}{3}}{y - 3} = \lim_{y \to 3} \frac{-y + 3}{3y(y - 3)} = \lim_{y \to 3} \frac{-1}{3y} = -\frac{1}{9}.$

13. $\displaystyle\lim_{z \to 8} \frac{z^{2/3}}{z - \sqrt{2z}} = \frac{\displaystyle\lim_{z \to 8} z^{2/3}}{\displaystyle\lim_{z \to 8} (z - \sqrt{2z})} = \frac{8^{2/3}}{8 - \sqrt{16}} = \frac{4}{4} = 1.$

16. $\displaystyle\lim_{h \to 0} \frac{\dfrac{1}{2 + h} - \dfrac{1}{2}}{h} = \lim_{h \to 0} \frac{2 - (2 + h)}{2h(2 + h)} = \lim_{h \to 0} \frac{-1}{2(2 + h)} = -\frac{1}{4}.$

19. $\displaystyle\lim_{x \to 3} |1 - x| = \left| \lim_{x \to 3} (1 - x) \right| = |1 - 3| = 2.$

22. $\displaystyle\lim_{x \to 9} \frac{3 - \sqrt{x}}{9 - x} = \lim_{x \to 9} \frac{3 - \sqrt{x}}{9 - x} \cdot \frac{3 + \sqrt{x}}{3 + \sqrt{x}} = \lim_{x \to 9} \frac{9 - x}{(9 - x)(3 + \sqrt{x})} = \lim_{x \to 9} \frac{1}{3 + \sqrt{x}} = \frac{1}{6}.$

   Alternatively, $\displaystyle\lim_{x \to 9} \frac{3 - \sqrt{x}}{9 - x} = \lim_{x \to 9} \frac{3 - \sqrt{x}}{(3 - \sqrt{x})(3 + \sqrt{x})} = \lim_{x \to 9} \frac{1}{3 + \sqrt{x}} = \frac{1}{6}.$

25. $\displaystyle\lim_{x \to 4} \frac{x^2 - 16}{2 - \sqrt{x}} = \lim_{x \to 4} \frac{(x + 4)(x - 4)}{2 - \sqrt{x}} \cdot \frac{2 + \sqrt{x}}{2 + \sqrt{x}}$

   $\displaystyle = \lim_{x \to 4} \frac{(x - 4)(x + 4)(2 + \sqrt{x})}{4 - x}$

   $\displaystyle = \lim_{x \to 4} -(x + 4)(2 + \sqrt{x}) = -32.$

28. $(-3)^{6/3} = 9$

31. 9

34. -2

37. 13

40. Does not exist

43. $f'(x) = \lim\limits_{h \to 0} \dfrac{\dfrac{x+h}{2x+2h+1} - \dfrac{x}{2x+1}}{h}$

$= \lim\limits_{h \to 0} \dfrac{2x^2 + 2xh + x + h - 2x^2 - 2xh - x}{h(2x+2h+1)(2x+1)}$

$= \lim\limits_{h \to 0} \dfrac{1}{(2x+2h+1)(2x+1)} = \dfrac{1}{(2x+1)^2}.$

46. $\lim\limits_{h \to 0}\left(\dfrac{\dfrac{1}{\sqrt{x+h+4}} - \dfrac{1}{\sqrt{x+4}}}{h}\right) = \lim\limits_{h \to 0} \dfrac{\sqrt{x+4} - \sqrt{x+h+4}}{h\sqrt{x+h+4}\sqrt{x+4}}$

$= \lim\limits_{h \to 0} \dfrac{\sqrt{x+4} - \sqrt{x+h+4}}{h\sqrt{x+h+4}\sqrt{x+4}} \cdot \dfrac{\sqrt{x+4} + \sqrt{x+h+4}}{\sqrt{x+4} + \sqrt{x+h+4}}$

$= \lim\limits_{h \to 0} \dfrac{(x+4) - (x+h+4)}{h\sqrt{x+h+4}\sqrt{x+4}(\sqrt{x+4} + \sqrt{x+h+4})}$

$= \lim\limits_{h \to 0} \dfrac{-1}{\sqrt{x+h+4}\sqrt{x+4}(\sqrt{x+4} + \sqrt{x+h+4})}$

$= \dfrac{-1}{2(x+4)\sqrt{x+4}} = -\dfrac{1}{2}(x+4)^{-3/2}.$

49. If x is near the integer n, then $f(x) = n - 1$ for $x < n$, and $f(x) = n$ for $x > n$. So $f(x)$ has no limit as $x \to n$. On the other hand, if r is not an integer, then $n < r < n+1$ for some integer n, and $f(x) = n$ if $n < x < n+1$, so $f(x) \to n$ as $x \to r$. Answer: The limit exists exactly when the real number a is not an integer.

52. That $h(x)$ approaches zero "steadily" as $x \to 0$ should mean this: If $|r| \le |s|$, then $|(h(r)| \le |h(s)|$. But here, $0.001 < 0.01$ whereas $h(0.001) = 0.001 > 0 = h(0.01)$. So $h(x)$ does not approach zero steadily as $x \to 0$. But because $h(x) = x$ if $|x| < 0.01$, $h(x) \to 0$ as $x \to 0$.

Section 2.3

1. $\theta \cdot \dfrac{\theta}{\sin\theta} \to 0 \cdot 1 = 0$ as $\theta \to 0$.

4. $\dfrac{\tan\theta}{\theta} = \dfrac{\sin\theta}{\theta\cos\theta} \to 1$ as $\theta \to 0$.

7. Let $z = 5x$. Then $z \to 0$ as $x \to 0$, and $\dfrac{\sin 5x}{x} = \dfrac{5\sin z}{z} \to 5$.

10. Replace $1 - \cos 2x$ with $1 - (\cos^2 x - \sin^2 x) = 2\sin^2 x$ to see that the limit is zero.

13. Multiply numerator and denominator by $1 + \cos x$.

16. $\lim\limits_{\theta \to 0} \dfrac{\sin 2\theta}{\theta} = \lim\limits_{\theta \to 0} \dfrac{2\sin\theta\cos\theta}{\theta} = 2 \cdot 1 \cdot 1 = 2.$

19. $\lim\limits_{x \to 0} \dfrac{\tan x}{x} = \lim\limits_{x \to 0} \dfrac{\sin x}{x} \cdot \dfrac{1}{\cos x} = 1 \cdot 1 = 1.$

22. $\lim\limits_{x \to 0} \dfrac{x - \tan x}{\sin x} = \lim\limits_{x \to 0} \dfrac{\left(\dfrac{x}{x} - \dfrac{\tan x}{x}\right)}{\left(\dfrac{\sin x}{x}\right)} = \dfrac{1 - 1}{1} = 0.$

25. $\lim\limits_{x \to 0} x^2 \csc 2x \cot 2x = \lim\limits_{x \to 0} \dfrac{1}{4} \cdot \dfrac{2x}{\sin 2x} \cdot \dfrac{2x}{\sin 2x} \cdot \cos 2x = \dfrac{1}{4} \cdot 1 \cdot 1 = \dfrac{1}{4}.$

28. Because $-1 \le \sin \dfrac{1}{x} \le 1$ for all $x \ne 0$, $-|\sqrt[3]{x}| \le \sqrt[3]{x}\sin \dfrac{1}{x} \le |\sqrt[3]{x}|$ for all $x \ne 0$. Now let $x \to 0$ to obtain the limit zero.

31. If $x < 1$, then $x - 1 < 0$, so the limit does not exist.

34. If $x > 3$ then $9 - x^2 < 0$, so the limit does not exist.

37. $\dfrac{4x}{x-4} \to +\infty$ as $x \to 4^+$, so the limit does not exist. In such a case it is also correct to write

$$\lim_{x \to 4^-} \sqrt{\dfrac{4x}{x-4}} = +\infty.$$

40. If $0 > x > -4$, then $16 - x^2 > 0$, so $\dfrac{16 - x^2}{\sqrt{16 - x^2}} = \sqrt{16 - x^2} \to 0$ as $x \to -4^+$.

43. If $x > 2$ then $x - 2 > 0$, so $\dfrac{2 - x}{|x - 2|} = -1$. Therefore the limit is also -1.

46. If $x < 0$ then $x - |x| = 2x$, so the limit is $1/2$.

49. $f(x) \to +\infty$ as $x \to 1^+$, $f(x) \to -\infty$ as $x \to 1^-$.

52. If x is near 5 then $2x - 5$ is near 5. So $f(x) \to +\infty$ as $x \to 5^-$, $f(x) \to -\infty$ as $x \to 5^+$.

55. If $x > 1$ then $f(x) = \dfrac{1}{x-1}$; if $x < 1$ then $f(x) = \dfrac{1}{1-x}$. Therefore $f(x) \to +\infty$ as $x \to 1$.

58. $\dfrac{x-1}{x^2 - 3x + 2} = \dfrac{x-1}{(x-1)(x-2)} = \dfrac{1}{x-2}$ for $x \ne 1$, $x \ne 2$. So $f(x) \to -1$ as $x \to 1$, $f(x) \to +\infty$ as $x \to 2^+$, and $f(x) \to -\infty$ as $x \to 2^-$.

61. If $2n - 1 < x < 2n$ then $f(x) = -1$, so $f(x) \to -1$ as $x \to 2n^-$. If $2n < x < 2n + 1$ then $f(x) = 1$, so $f(x) \to 1$ as $x \to 2n^+$. A similar analysis will yield the answers for limits of f at odd integers.

64. If $x > 0$, then $\dfrac{f(x+h) - f(x)}{h} = \dfrac{x + h - x}{h} = 1$ because $x + h > 0$ for h sufficiently near zero. Hence $\dfrac{f(x+h) - f(x)}{h} \to 1$ as $h \to 0$ in the case $x > 0$. The case $x < 0$, in which you will find that $f'(x) \equiv -1$, is similar.

Section 2.4

1. $f(g(x)) = 1 - (2x + 3)^2 = -4x^2 - 12x - 8$; $\quad g(f(x)) = 2(1 - x^2) + 3 = -2x^2 + 5$.

4. $f(g(x)) = \left(\dfrac{1}{x^2 + 1}\right)^2 + 1$; $\quad g(f(x)) = \dfrac{1}{(x^2 + 1)^2 + 1}$.

7. $f(g(x)) = \sin(g(x)) = \sin(x^3) = \sin x^3$; $\quad g(f(x)) = g(\sin x) = (\sin x)^3 = \sin^3 x$. The graphs of $f(g)$ and $g(f)$ are shown next. Which is which?

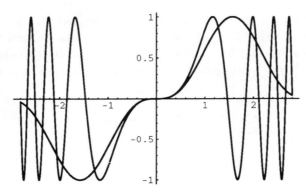

10. $f(g(x)) = 1 - \sin^2 x = \cos^2 x$; $\quad g(f(x)) = \sin(1 - x^2)$.

13. $h(x) = (2x - x^2)^{1/2}$: $k = 1/2$, $g(x) = 2x - x^2$

16. $h(x) = (4x - 6)^{4/3} = [(4x - 6)^4]^{1/3} = [(4x - 6)^{1/3}]^4$, so here are three correct answers (and there are others): $k = 4/3$ and $g(x) = 4x - 6$; $k = 1/3$ and $g(x) = (4x - 6)^4$; $k = 4$ and $g(x) = (4x - 6)^{1/3}$.

19. $k = 1/2$ and $g(x) = \dfrac{1}{\sqrt{x + 10}}$; $k = -1/2$ and $g(x) = x + 10$.

22. The domain is the set of all real numbers other than zero. The function f is continuous on its domain (but is not continuous at $x = 0$ because it is not defined there).

25. Continuous on its domain $\mathbb{R}$ (the denominator is never zero).

28. Continuous on $\mathbb{R}$ (the denominator is never zero).

31. Every real number has a [unique] cube root, so by Theorem 2 and the limit laws f is continuous on its domain, $x \neq 1$.

34. Continuous on its domain, the interval $[-3, 3]$. This follows from Theorem 2 and the limit laws.

37. The function f is continuous on its domain, the set of all real numbers other than zero.

40. The domain of f consists of those values of x for which $\sin x \geq 0$. So its domain is the union of all intervals of the form $[n\pi, (n + 1)\pi]$ where n is an even integer. It is continuous on its domain by Theorem 2.

43. The function f is discontinuous at $x = -3$, and cannot be made continuous there because it has no limit there.

46. $f(x) = \dfrac{x + 1}{x^2 - x - 6} = \dfrac{x + 1}{(x - 3)(x + 2)}$, so f has no limit at either $x = 3$ or $x = -2$. Therefore it is not continuous at either point nor can it be made continuous at either point.

49. Because $f(x) \to 1$ as $x \to 17^+$ whereas $f(x) \to -1$ as $x \to 17^-$, f has no limit at $x = 17$ and cannot be made continuous there.

52. The function f is not continuous at $x = 1$ because it is not defined there. If we define $f(1)$ to be 2, then f will become continuous at $x = 1$ as well because its limit at $x = 1$ will be equal to its value there.

55. Let $f(x) = x^3 - 3x^2 + 1$. Then f is continuous on $[0, 1]$ because $f(x)$ is a polynomial. Moreover, $f(0) = 1 > 0 > -1 = f(1)$. Therefore $f(x) = 0$ for some number x in $[0, 1]$ ($x \approx 0.6527$).

58. Let $f(x) = x^5 - 5x^3 + 3$. Then f is continuous on $[-3, -2]$, because $f(x)$ is a polynomial. Moreover, $f(-3) = -105 < 0$, and $f(-2) = 11 > 0$. Therefore $f(x) = 0$ for some number x in $[-3, -2]$. (The value of that number x is approximately -2.291164.)

61. Given $a > 0$, let $f(x) = x^2 - a$. Then $f(0) = -a < 0$ and $f(a + 1) = a^2 + a + 1 > 0$. Because $f(x)$ is a polynomial, f is continuous on $[0, a + 1]$, so f has the intermediate value property there. Therefore $f(r) = 0$ for some number r in $(0, a + 1)$. That is, $r^2 = a$, and therefore every positive real number has a square root.

64. If r is not an integer, then $n < r < n + 1$ where n is an integer, and $f(x) = x + n$ for all x near r. So f is continuous at non-integers. If r is an integer, then $f(x)$ is near $2r$ for $x > r$ and close to r, but $f(x)$ is near $2r - 1$ for $x < r$ and close to r. So the left-hand and right-hand limits of f are unequal at integers. Therefore f is discontinuous at every integer and continuous at every non-integer.

Chapter 2 Miscellaneous

1. $0^2 - 3 \cdot 0 + 4 = 4$

4. $(1 + 1 - 1)^{17} = 1$

7. $\dfrac{x^2 - 1}{1 - x} = -(x + 1) \to -2$ as $x \to 1$.

10. $\dfrac{4 - x^2}{3 + x} \to \dfrac{4}{3}$ as $x \to 0$.

13. $16^{3/4} = 8$

16. $x - \sqrt{x^2 - 1} = \dfrac{x^2 - (x^2 - 1)}{x + \sqrt{x^2 - 1}} = \dfrac{1}{x + \sqrt{x^2 - 1}} \to 1$ as $x \to 1^+$.

19. $\sqrt{(2 - x)^2} = |2 - x| = x - 2$ for $x > 2$, so the fraction and its limit are equal to -1.

22. If $x < 3$ then $x^2 - 9 < 0$, so the limit does not exist.

25. The numerator approaches 4 while the denominator approaches zero through positive values. Therefore the limit is $+\infty$.

28. If $x \neq 1$ then the fraction is equal to $\dfrac{1}{x - 1}$, and its denominator approaches zero through negative values, so the limit is $-\infty$.

31. $\dfrac{\sin 3x}{x} = \dfrac{3 \sin 3x}{3x} \to 3$ as $x \to 0$.

34. $\lim\limits_{x \to 0} \dfrac{\tan 2x}{\tan 3x} = \lim\limits_{x \to 0} \dfrac{2}{3} \cdot \dfrac{\sin 2x}{2x} \cdot \dfrac{3x}{\sin 3x} \cdot \dfrac{\cos 3x}{\cos 2x} = \dfrac{2}{3}$.

37. $\lim\limits_{x \to 0} \dfrac{1 - \cos 3x}{2x^2} = \lim\limits_{x \to 0} \dfrac{1 - \cos^2 3x}{2x^2(1 + \cos 3x)} = \lim\limits_{x \to 0} \dfrac{9}{2(1 + \cos 3x)} \cdot \dfrac{\sin 3x}{3x} \cdot \dfrac{\sin 3x}{3x} = \dfrac{9}{4}$.

40. $\lim\limits_{x \to 0} x^2 \cot^2 3x = \lim\limits_{x \to 0} \dfrac{3x}{\sin 3x} \cdot \dfrac{3x}{\sin 3x} \cdot \dfrac{\cos^2 3x}{9} = \dfrac{1}{9}$.

43. $f'(x) = 6x + 4$; $f'(1) = 10$: $f(1) = 2$; $y - 2 = 10(x - 1)$.

46. $f(x) = \frac{1}{3}x - \frac{1}{16}x^2$, so $f'(x) = \frac{1}{3} - \frac{1}{8}x$. Then $f(1) = \frac{13}{48}$ and $f'(1) = \frac{5}{24}$, so one equation of the line is $y - \frac{13}{48} = \frac{5}{24}(x - 1)$.

49. $f'(x) = \lim\limits_{h \to 0} \dfrac{\dfrac{1}{3 - x - h} - \dfrac{1}{3 - x}}{h} = \lim\limits_{h \to 0} \dfrac{3 - x - 3 + x + h}{h(3 - x - h)(3 - x)}$

$= \lim\limits_{h \to 0} \dfrac{1}{(3 - x - h)(3 - x)} = \dfrac{1}{(3 - x)^2}$.

52. $f'(x) = \lim\limits_{h \to 0} \dfrac{\dfrac{x + h}{x + h + 1} - \dfrac{x}{x + 1}}{h} = \lim\limits_{h \to 0} \dfrac{x^2 + xh + x + h - x^2 - xh - x}{h(x + h + 1)(x + 1)}$

$= \lim\limits_{h \to 0} \dfrac{1}{(x + h + 1)(x + 1)} = \dfrac{1}{(x + 1)^2}$.

55. The slope of a line tangent to $y = x^2$ at (a, a^2) is $2a$. If such a line passes through $(3, 4)$, then the two-point slope formula yields $2a = \dfrac{a^2 - 4}{a - 3}$. It is now easy to solve to obtain the two values $a = 3 \pm \sqrt{5}$.

58. $(1 + \sqrt{x})^2 = 1 + 2\sqrt{x} + x$; $2 + (1 + \sqrt{x})^2 = 3 + 2\sqrt{x} + x$; $f(x) = 2 + x^2$.

64. Every rational function is continuous wherever its denominator is nonzero. Here the denominator of $f(x)$ is zero when $x = 2$, and f cannot be made continuous there.

67. Let $f(x) = x^5 + x - 1$. Then $f(0) = -1 < 0 < 1 = f(1)$. Because $f(x)$ is a polynomial, it is continuous on $[0, 1]$, so f has the intermediate value property there. Hence there exists a number c in $(0, 1)$ such that $f(c) = 0$. Thus $c^5 + c - 1 = 0$, and so the equation $x^5 + x - 1 = 0$ has a solution. (The value of c is approximately 0.754877666.)

70. Let $h(x) = x + \tan x$. Then $h(\pi) = \pi > 0$ and $\lim\limits_{x \to (\pi/2)^+} h(x) = -\infty$. The latter fact implies that $h(r) < 0$ for some number r slightly larger than $\pi/2$. Because h is continuous on the interval $[r, \pi]$, h has the intermediate value property there, so $h(c) = 0$ for some number c between r and π, and thus between $\pi/2$ and π.

1. $f'(x) = 4 \qquad (a = 0, \ b = 4, \ c = -5)$

4. $f'(x) = -49$

10. $\dfrac{du}{dt} = 14t + 13$

13. $f'(x) = \lim\limits_{h \to 0} \dfrac{(x+h)^2 + 5 - x^2 - 5}{h} = \lim\limits_{h \to 0} \dfrac{2xh + h^2}{h} = \lim\limits_{h \to 0} (2x + h) = 2x.$

16. $f'(x) = \lim\limits_{h \to 0} \dfrac{\dfrac{1}{3 - x - h} - \dfrac{1}{3 - x}}{h} = \lim\limits_{h \to 0} \dfrac{3 - x - 3 + x + h}{h(3 - x - h)(3 - x)}$

$\qquad = \lim\limits_{h \to 0} \dfrac{1}{(3 - x - h)(3 - x)} = \dfrac{1}{(3 - x)^2}.$

19. $f'(x) = \lim\limits_{h \to 0} \dfrac{\dfrac{x + h}{1 - 2x - 2h} - \dfrac{x}{1 - 2x}}{h} = \lim\limits_{h \to 0} \dfrac{(x + h)(1 - 2x) - x(1 - 2x - 2h)}{h(1 - 2x - 2h)(1 - 2x)}$

$\qquad = \lim\limits_{h \to 0} \dfrac{1}{(1 - 2x - 2h)(1 - 2x)} = \dfrac{1}{(1 - 2x)^2}.$

22. $\dfrac{dx}{dt} = -32t + 160; \quad \dfrac{dx}{dt} = 0$ when $t = 5; \quad x(5) = 425$

25. $\dfrac{dx}{dt} = -20 - 10t; \quad \dfrac{dx}{dt} = 0$ when $t = -2; \quad x(-2) = 120$

28. $\dfrac{dy}{dt} = -32t + 128; \quad \dfrac{dy}{dt} = 0$ when $t = 4; \quad y(4) = 281$ (ft)

31. Let r denote the radius of the circle. Then $A = \pi r^2$ and $C = 2\pi r$, so $r = \dfrac{C}{2\pi}$. Thus $A = \dfrac{1}{4\pi}C^2$, so the rate of change of A with respect to C is $dA/dC = \dfrac{C}{2\pi}$.

34. Because $V(t) = 10 - \frac{1}{5}t + \frac{1}{1000}t^2$, $V'(t) = -\frac{1}{5} + \frac{1}{500}t$ and the rate at which the water is leaking out one minute later ($t = 60$) is $V(60) = -\frac{2}{25}$ (gal/s) or—if you prefer— -4.8 gal/min. The average rate of change of V from $t = 0$ until $t = 100$ is $\dfrac{V(100) - V(0)}{100 - 0} = \dfrac{0 - 10}{100} = -\dfrac{1}{10}$. The instantaneous rate of change of V will have this value when $V'(t) = -\frac{1}{10}$, which we easily solve for $t = 50$.

37. On our graph, the tangent line at the point $(20, 810)$ has slope $m_1 \approx 0.6$ and the tangent line at $(40, 2686)$ has slope $m_2 \approx 0.9$. A line of slope 1 on our graph corresponds to a velocity of 125 ft/s (because the line through $(0, 0)$ and $(10, 1250)$ has slope 1), and thus we estimate the velocity of the car at time $t = 20$ to be about $(0.6)(125) = 75$ ft/s, and at time $t = 40$ it is traveling at about $(0.9)(125) = 112.5$ ft/s. The method is crude; the answer in the back of the textbook is quite different simply because it was obtained by someone else.

40. A right circular cylinder of radius r and height h has volume $V = \pi r^2 h$ and total surface area S obtained by adding the areas of its top, bottom, and curved side: $S = 2\pi r^2 + 2\pi r h$. We are given $h = 2r$, so $V(r) = 2\pi r^3$ and $S(r) = 6\pi r^2$. Also $dV/dr = 6\pi r^2 = S(r)$, so the rate of change of volume with respect to radius is indeed equal to total surface area.

43. Let $V(t)$ denote the volume (in cm^3) of the snowball at time t (in hours), and let $r(t)$ denote its radius then. From the data given in the problem, $r = 12 - t$. The volume of the snowball is

$$V = \frac{4}{3}\pi r^3 = \frac{4}{3}\pi(12 - t)^3 = \frac{4}{3}\pi(1728 - 432t + 36t^2 - t^3),$$

so its instantaneous rate of change is

$$V'(t) = \frac{4}{3}\pi(-432 + 72t - 3t^2).$$

Hence its rate of change of volume when $t = 6$ is $V'(6) = -144\pi$ cm^3/h. Its average rate of change of volume from $t = 3$ to $t = 9$ in cm^3/h is

$$\frac{V(9) - V(3)}{9 - 3} = \frac{36\pi - 972\pi}{6} = -156\pi.$$

46.   Because $P(t) = 100 + 4t + \frac{3}{10}t^2$, we have $P'(t) = 4 + \frac{3}{5}t$. The year 1986 corresponds to $t = 6$, so the rate of change of P then was $P'(6) = 7.6$ (thousands per year). The average rate of change of P from 1983 ($t = 3$) to 1988 ($t = 8$) was

$$\frac{P(8) - P(3)}{8 - 3} = \frac{151.2 - 114.7}{5} = 7.3 \text{ (thousands per year)}.$$

Section 3.2

1. $f'(x) = 6x - 1$

4. $g'(x) = (2x^2 - 1)(3x^2) + (4x)(x^3 + 2) = 10x^4 - 3x^2 + 8x$

7. $f(y) = 4y^3 - y$: $f'(y) = 12y^2 - 1$.

10. $f'(t) = \dfrac{0 - (1)(-2t)}{(4 - t^2)^2} = \dfrac{2t}{(4 - t^2)^2}$

13.   By first multiplying the two factors of $g(t)$:

$$g'(t) = 5t^4 + 4t^3 + 3t^2 + 4t.$$

By applying the product rule:

$$g'(t) = (t^2 + 1)(3t^2 + 2t) + (t^3 + t^2 + 1)(2t).$$

Of course both answers are the same.

16. $f(x) = 2x - 3 + 4x^{-1} - 5x^{-2}$, so $f'(x) = 2 - \dfrac{4}{x^2} + \dfrac{10}{x^3} = \dfrac{2x^3 - 4x + 10}{x^3}$.

22. $h'(x) = \dfrac{(2x - 5)(6x^2 + 2x - 3) - (2)(2x^3 + x^2 - 3x + 17)}{(2x - 5)^2} = \dfrac{8x^3 - 28x^2 - 10x - 19}{(2x - 5)^2}$

25.   First rewrite the function in the form $g(x) = \dfrac{x^3 - 2x^2}{2x - 3}$.

Then $g'(x) = \dfrac{(2x - 3)(3x^2 - 4x) - (2)(x^3 - 2x^2)}{(2x - 3)^2} = \dfrac{4x^3 - 13x^2 + 12x}{(2x - 3)^2}$.

28. $\dfrac{dy}{dx} = -3x^{-2} + 8x^{-3}$

34. $y = \dfrac{x^2}{x^2 - 4}$ for $x \neq 0$. So $\dfrac{dy}{dx} = \dfrac{(x^2 - 4)(2x) - (x^2)(2x)}{(x^2 - 4)^2} = -\dfrac{8x}{(x^2 - 4)^2}$.

37. $y = \dfrac{10x^6}{15x^5 - 4}$ for $x \neq 0$.

40. $y = x^{-1} + 10x^{-2}$, so $\dfrac{dy}{dx} = -x^{-2} - 20x^{-3}$.

43. $\dfrac{dy}{dx} = -(x - 1)^2$; the slope at P is -1. An equation of the tangent line is $y - 1 = -(x - 2)$.

46. $y = \dfrac{x^2}{x - 1}$ for $x \neq 0$, so $\dfrac{dy}{dx} = \dfrac{(x - 1)(2x) - x^2}{(x - 1)^2} = \dfrac{x^2 - 2x}{(x - 1)^2}$. The slope at P is zero; the line has equation $y = 4$.

49. $\dfrac{dy}{dx} = \dfrac{3x^2 + 6x}{(x^2 + x + 1)^2}$

52. $W = \dfrac{2 \times 10^9}{R^2} = (2 \times 10^9)R^{-2}$, so $\dfrac{dW}{dR} = -\dfrac{4 \times 10^9}{R^3}$; when $R = 3960$, $\dfrac{dW}{dR} = -\dfrac{62500}{970299}$ lb/mi. Thus W decreases initially at about 1.03 oz/mi.

55. The slope of the tangent line can be computed using dy/dx at $x = a$ and also by using the two points known to lie on the line. We thereby find that $3a^2 = \dfrac{a^3 - 5}{a - 1}$. This leads to the equation $(a + 1)(2a^2 - 5a + 5) = 0$. The quadratic factor has negative discriminant, so the only real solution of the cubic equation is $a = -1$. The point of tangency is $(-1, -1)$, the slope there is 3, and the equation of the line in question is $y = 3x + 2$.

58. Let $(a, 1/a)$ be a point of tangency. The slope of the tangent there is $-1/a^2$, so $-1/a^2 = -2$. Thus there are two possible values for a: $\pm\frac{1}{2}\sqrt{2}$. These lead to the equations of the two lines: $y = -2x + 2\sqrt{2}$ and $y = -2x - 2\sqrt{2}$.

61. $D_x[f(x)]^3 = f'(x)f(x)f(x) + f(x)f'(x)f(x) + f(x)f(x)f'(x) = 3[f(x)]^2 f'(x)$.

64. With $f(x) = x^2 + x + 1$ and $n = 100$, we obtain $D_x(x^2 + x + 1)^{100} = 100(x^2 + x + 1)^{99}(2x + 1)$.

67. When $n = 0$, $\dfrac{dy}{dx} = -\dfrac{2x}{(1 + x^2)^2}$; when $n = 2$, $\dfrac{dy}{dx} = \dfrac{2x}{(1 + x^2)^2}$. In both cases there is only one horizontal tangent, at the point where $x = 0$.

70. $f'(x) = 1$ when $x = \pm 1$.

Section 3.3

1. $dy/dx = 5(3x + 4)^4(3) = 15(3x + 4)^4$

4. $y = (2x + 1)^{-3}$: $dy/dx = -6(2x + 1)^{-4}$

7. $dy/dx = -4(2 - x)^3(3 + x)^7 + 7(2 - x)^4(3 + x)^6$

10. $dy/dx = \dfrac{-6x(4 + 5x + 6x^2)(1 - x^2)^2 - 2(5 + 12x)(1 - x^2)^3}{(4 + 5x + 6x^2)^3}$.
    A common factor of $4 + 5x + 6x^2$ was cancelled from both terms in the numerator and from the denominator.

13. $dy/dx = 3(u + 1)^2 du/dx = -\dfrac{6}{x^3}\left(\dfrac{1}{x^2} + 1\right)^2$

16. $dy/dx = \dfrac{-15}{(3x - 2)^6}$

19. $dy/dx = -\dfrac{2(x^3 - 1)^2(x^3 + 1)^2(5x^6 - 14)}{x^{29}}$. This is the result of considerable simplification of the initial result, which was $dy/dx = -4x^{-5}(x^{-2} - x^{-8})^3 + 3x^{-4}(8x^{-9} - 2x^{-3})(x^{-2} - x^{-8})^2$.

22. $f'(x) = -\dfrac{15x^2}{(5x^3 + 2)^2}$

28. $h'(z) = 6z^3(z^2 + 4)^2 + 2z(z^2 + 4)^3$

31. $f'(u) = 8u(u + 1)^3(u^2 + 1)^3 + 3(u + 1)^2(u^2 + 1)^4$

34. $p'(t) = \dfrac{4(t^{-2} + 2t^{-3} + 3t^{-4})}{(t^{-1} + t^{-2} + t^{-3})^5} = \dfrac{4t^{11}(t^2 + 2t + 3)}{(t^2 + t + 1)^5}$

40. $dy/dx = -3(1 - x)^2$

43. $dy/dx = -\dfrac{2x}{(x^2 + 1)^2}$

46. $g'(t) = 3\sin^2 t \cos t$

49. If the circle has area A and radius r, then $A = \pi r^2$. If t denotes time (in seconds), then we are given

$dr/dt = 2$ at the time in question. But

$$\frac{dA}{dt} = \frac{dA}{dr} \cdot \frac{dr}{dt} = 2\pi r \cdot 2 = 4\pi r,$$

so when $r = 10$ (in.), the area of the circle is increasing at 40π in.^{2}/s.

52. Let x denote the length of each side of the triangle. Then its altitude is $\frac{1}{2}x\sqrt{3}$, and so its area is $A = \frac{1}{4}x^2\sqrt{3}$. Therefore the rate of change of its area with respect to time t (in seconds) is

$$\frac{dA}{dt} = \frac{1}{2}x\sqrt{3}\,\frac{dx}{dt}.$$

We are given $x = 10$ and $dx/dt = 2$, so at that point the area is increasing at $10\sqrt{3}$ in.^{2}/s.

55. $G'(t) = f'(h(t)) \cdot h'(t)$. Now $h(1) = 4$, $h'(1) = -6$, and $f'(4) = 3$, so $G'(1) = 3 \cdot (-6) = -18$.

58. Let V denote the volume of the balloon and r its radius at time t (in seconds). We are given $dV/dt = 200\pi$. Now

$$\frac{dV}{dt} = \frac{dV}{dr} \cdot \frac{dr}{dt} = 4\pi r^2 \frac{dr}{dt}.$$

When $r = 5$, we have $200\pi = 4\pi \cdot 25 \cdot (dr/dt)$, so $dr/dt = 2$. Answer: When $r = 5$, (cm), the radius of the balloon is increasing at 2 cm/s.

61. Let V denote the volume of the snowball and A its surface area at time t (in hours). Then

$$dV/dt = kA \quad \text{and} \quad A = cV^{2/3}$$

(the latter because A is proportional to r^2, whereas V is proportional to r^3). Therefore

$$dV/dt = \alpha V^{2/3} \quad \text{and thus} \quad dt/dV = \beta V^{-2/3}$$

(α and β are constants). From the last equation we may conclude that $t = \gamma V^{1/3} + \delta$ for some constants γ and δ, so that $V = V(t) = (Pt + Q)^3$ for some constants P and Q. From the information $500 = V(0) = Q^3$ and $250 = V(1) = (P + Q)^3$, we find that $Q = 5\sqrt[3]{4}$ and that $P = -5 \cdot (\sqrt[3]{4} - \sqrt[3]{2})$. Now $V(t) = 0$ when $PT + Q = 0$; it turns out that

$$T = \frac{\sqrt[3]{2}}{\sqrt[3]{2} - 1} \approx 4.8473.$$

Therefore the snowball finishes melting at about 2:50:50 P.M. on the same day.

Section 3.4

1. $f(x) = 4x^{5/2} + 2x^{-1/2}$, so $f'(x) = 10x^{3/2} - x^{-3/2}$.

4. $f(x) = (7 - 6x)^{-1/3}$, so $f'(x) = 2(7 - 6x)^{-4/3}$.

7. $f'(x) = \frac{3}{2}(2x + 3)^{1/2} \cdot 2 = 3(2x + 3)^{1/2}$

10. $f'(x) = -\frac{2}{3}(4 - 3x^3)^{-5/3} \cdot (-9x) = 6x(4 - 3x^3)^{-5/3}$

13. $f(x) = (2x^2 + 1)^{1/2}$, so $f'(x) = 2x(2x^2 + 1)^{-1/2}$.

16. $g(t) = (3t^5)^{-1/2}$, so $g'(t) = -\frac{1}{2}(3t^5)^{-3/2} \cdot 15t^4$.

19. $g(x) = (x - 2x^3)^{-4/3}$, so $g'(x) = -\frac{4}{3}(x - 2x^3)^{-7/3}(1 - 6x^2)$.

22. $g(x) = \left(\dfrac{2x + 1}{x - 1}\right)^{1/2}$, so $g'(x) = \dfrac{1}{2}\left(\dfrac{2x + 1}{x - 1}\right)^{-1/2} \cdot \dfrac{(x - 1) \cdot 2 - (2x + 1) \cdot 1}{(x - 1)^2} = \dfrac{-3}{2(x - 1)^2}\sqrt{\dfrac{x - 1}{2x + 1}}$.

25. $f'(x) = 3\left(x - \dfrac{1}{x}\right)^2\left(1 + \dfrac{1}{x^2}\right)$

28. $h'(x) = \dfrac{5}{3}\left(\dfrac{x}{1+x^2}\right)^{2/3} \cdot \dfrac{1-x^2}{(1+x^2)^2}$

31. $f'(x) = (3-4x)^{1/2} - 2x(3-4x)^{-1/2}$

34. $f'(x) = -\frac{1}{2}(1-x)^{-1/2}(2-x)^{1/3} - \frac{1}{3}(1-x)^{1/2}(2-x)^{-2/3}$

37. $f'(x) = \dfrac{2(3x+4)^5 - 15(3x+4)^4(2x-1)}{(3x+4)^{10}}$, which can be simplified to $f'(x) = \dfrac{23-24x}{(3x+4)^6}$.

40. $f'(x) = -5(1-3x^4)^4(12x^3)(4-x)^{1/3} - \frac{1}{3}(1-3x^4)^5(4-x)^{-2/3}$

43. $g'(t) = \frac{1}{2}[t+(t+t^{1/2})^{1/2}]^{-1/2}[1 + \frac{1}{2}(t+t^{1/2})^{-1/2}(1+\frac{1}{2}t^{-1/2})] = \dfrac{1+\dfrac{1+\dfrac{1}{2\sqrt{t}}}{2\sqrt{t+\sqrt{t}}}}{2\sqrt{t+\sqrt{t+\sqrt{t}}}}$

46. $\dfrac{dy}{dx} = \dfrac{4-2x^2}{\sqrt{4-x^2}}$, so there are horizontal tangents at $(\sqrt{2},2)$ and $(-\sqrt{2},-2)$. There are vertical tangents at $(2,0)$ and $(-2,0)$.

49. $\dfrac{dy}{dx} = \dfrac{1}{(1-x^2)^{3/2}}$, so there are no horizontal tangents. There are no vertical tangents because $y(x)$ is undefined at $x = \pm 1$.

52. $dV/dS = \frac{1}{4}\sqrt{S/\pi}$, and $S = 400\pi$ when the radius of the sphere is 10, so the answer is 5 (in appropriate units).

55. The line tangent to the parabola $y = x^2$ at the point $Q(a,a^2)$ has slope $2a$, so the normal to the parabola at Q has slope $-1/(2a)$. The normal also passes through $P(18,0)$, so we can find its slope another way—by using the two-point formula. Thus

$$-\frac{1}{2a} = \frac{a^2-0}{a-18};$$

$$18 - a = 2a^3;$$

$$2a^3 + a - 18 = 0.$$

By inspection, $a = 2$ is a solution. Thus $a - 2$ is a factor of the cubic, so

$$2a^3 + a - 18 = (a-2)(2a^2 + 4a + 9).$$

The quadratic factor has negative discriminant, so $a = 2$ is the only real solution of $2a^3 + a - 18 = 0$. Therefore the normal line has slope $-1/4$ and equation $x + 4y = 18$.

58. Equation (3) is an *identity*, and if two functions have identical graphs on an interval, then their derivatives will also be identically equal to each other on that interval. There is no point in differentiating both sides of an algebraic equation.

Section 3.5

1. Maximum value 2 at $x = -1$ because f is a decreasing function; no minimum value because $(1,0)$ is not on the graph.

4. No maximum value because $\lim\limits_{x \to 0^+} f(x) = +\infty$. Minimum value 1 at $x = 1$ because f is a decreasing function.

7. The graph of f is increasing, so f has a minimum at $(-1,0)$ and a maximum at $(1,2)$.

10. Minimum 4 at $x = 1/2$, no maximum because $f(x) \to +\infty$ as $x \to 1^-$ and as $x \to 0^+$.

13. Because $h'(x) = -2x$ is never zero on the domain $[1, 3]$ of h, the extrema can occur only at the endpoints of the domain. And $h(3) = -5 < 3 < h(1)$, so the minimum value of h is -5 and its maximum value is 3.

16. $h'(x) = 2x + 4$; $h'(x) = 0$ when $x = -2$. $h(-3) = 4$, $h(-2) = 3$ (minimum), and $h(0) = 7$ (maximum).

19. $h'(x) = 1 - \dfrac{4}{x^2}$; $h'(x) = 0$ when $x = \pm 2$, but -2 is not in the domain of h. The minimum is $h(2) = 4$ and the maximum value of h is $h(1) = h(4) = 5$.

22. $f'(x) = 2x - 4$; $f'(x) = 0$ when $x = 2$. $f(0) = 3$ (maximum) and $f(2) = -1$ (minimum).

25. $f'(x) = 3x^2 - 6x - 9 = 3(x + 1)(x - 3)$; $f'(x) = 0$ when $x = -1$ and when $x = 3$. $f(-2) = 3$, $f(-1) = 10$ (maximum), $f(3) = -22$ (minimum), and $f(4) = -15$.

28. $f'(x) = -2$ for $1 < x < 3/2$, $f'(x) = 2$ for $3/2 < x < 2$; $f'(3/2)$ does not exist. $f(1) = 1 = f(2)$ (maximum) and $f(3/2) = 0$ (minimum).

31. $f'(x) = 150x^2 - 210x + 72 = 6(5x - 3)(5x - 4)$; $f'(x) = 0$ when $x = 3/5$ and when $x = 4/5$. $f(0) = 0$ (minimum), $f(3/5) = 16.2$, $f(4/5) = 16$, and $f(1) = 17$ (maximum).

34. $f'(x) = \dfrac{1 - x^2}{(1 + x^2)^2}$; $f'(x)$ always exists and $f'(x) = 0$ when $x = 1$ (and when $x = -1$, but the latter is not in the domain of f). $f(0) = 0$ (minimum), $f(1) = 1/2$ (maximum), and $f(3) = 0.3$.

37. $f'(x) = \dfrac{1 - 2x^2}{(1 - x^2)^{1/2}}$; $f'(x) = 0$ when $x = \pm\sqrt{2}/2$; $f'(x)$ does not exist when $x = \pm 1$ (the endpoints of the domain of f). $f(-1) = 0$, $f(-\sqrt{2}/2) = -1/2$ (minimum), $f(\sqrt{2}/2) = 1/2$ (maximum), and $f(1) = 0$.

40. $f'(x) = \dfrac{1 - 3x}{2\sqrt{x}}$; $f'(x) = 0$ when $x = 1/3$, and $f'(0)$ does not exist. $f(0) = 0$, $f(1/3) = 2\sqrt{3}/9$ (maximum), and $f(4) = -6$ (minimum).

43. $f'(x) = 0$ if x is not an integer; $f'(x)$ does not exist if x is an integer.

46. Every real number that is an integer or half an odd integer is a critical point of f; f' does not exist at each such real number.

49. (d)

52. (e)

Section 3.6

1. With $x > 0$, $y > 0$, and $x + y = 50$, we are to maximize the product $P = xy$.

$$P = P(x) = x(50 - x) = 50x - x^2, \qquad 0 < x < 50$$

($x < 50$ because $y > 0$.) The product is not maximal if we let $x = 0$ or $x = 50$, so we adjoin the endpoints to the domain of P; thus the continuous function $P(x) = 50x - x^2$ has a global maximum on the closed interval $[0, 50]$, and the maximum does *not* occur at either endpoint. Because f is differentiable, the maximum must occur at a point where $P'(x) = 0$: $50 - 2x = 0$, and so $x = 25$. Because this is the only critical point of P, it follows that $x = 25$ maximizes $P(x)$. When $x = 25$, $y = 50 - 25 = 25$, so the two positive real numbers with sum 50 and maximum possible product are 25 and 25.

4. If the side of the pen parallel to the wall has length x and the two perpendicular sides both have length y, then we are to maximize area $A = xy$ given $x + 2y = 600$. Thus

$$A = A(y) = y(600 - 2y), \qquad 0 \le y \le 300.$$

Adjoining the endpoints to the domain is allowed because the maximum we seek occurs at neither endpoint. Therefore the maximum occurs at an interior critical point. We have $A'(y) = 600 - 4y$, so the only critical point of A is $y = 150$. Then $y = 150$, we have $x = 300$, so the maximum possible area that can be enclosed is 45,000 m^2.

7. If the two numbers are x and y, then we are to minimize $S = x^2 + y^2$ given $x > 0$, $y > 0$, and $x + y = 48$. So $S(x) = x^2 + (48 - x)^2$, $0 \leq x \leq 48$. Here we adjoin the endpoints to the domain to ensure the existence of a maximum, but we must test the values of S at these endpoints because it is not immediately clear that neither $S(0)$ nor $S(48)$ yields the maximum value of S. Now $S'(x) = 2x - 2(48 - x)$; the only interior critical point of S is $x = 24$, and when $x = 24$, $y = 24$ as well. Now $S(0) = (48)^2 = 2304 = S(48) > 1152 = S(24)$, so the answer is 1152.

10. Draw a cross section of the cylindrical log—a circle of radius r. Inscribe in this circle a cross section of the beam—a rectangle of width w and height h. Draw a diagonal of the rectangle; the Pythagorean theorem yields $x^2 + h^2 = 4r^2$. The strength S of the beam is given by $S = kwh^2$ where k is a positive constant. Because $h^2 = 4r^2 - w^2$, we have

$$S = S(w) = kw(4r^2 - w^2) = k(4wr^2 - w^3)$$

with natural domain $0 < w < 2r$. We adjoin the endpoints to this domain; this is permissible because $S = 0$ at each, and so is not maximal. Next, $S'(w) = k(4r^2 - 3w^2)$; $S'(w) = 0$ when $3w^2 = 4r^2$, and the corresponding (positive) value of w yields the maximum of S (we know that $S(w)$ must have a maximum on $[0, 2r]$ because of the continuity of S on this interval, and we also know that the maximum does not occur at either endpoint, so there is only one possible location for the maximum). At maximum, $h^2 = 4r^2 - w^2 = 3w^2 - w^2$, so $h = w\sqrt{2}$ describes the shape of the beam of greatest strength.

13. If the rectangle has sides x and y, then $x^2 + y^2 = 16^2$ by the Pythagorean theorem. The area of the rectangle is then

$$A(x) = 2\sqrt{256 - x^2}, \qquad 0 \leq x \leq 16.$$

A positive quantity is maximized exactly when its square is maximized, so in place of A we maximize

$$f(x) = (A(x))^2 = 256x^2 - x^4.$$

The only solutions of $f'(x) = 0$ in the domain of A are $x = 0$ and $x = 8\sqrt{2}$; the former minimizes $A(x)$ and the latter yields its maximum value, 128.

16. Let $P(x, 0)$ be the lower right-hand corner point of the rectangle. The rectangle then has base $2x$, height $4 - x^2$, and thus area

$$A(x) = 2x(4 - x^2) = 8x - 2x^3, \qquad 0 \leq x \leq 2.$$

Now $A'(x) = 8 - 6x^2$; $A'(x) = 0$ when $x = 2\sqrt{3}/3$. Because $A(0) = 0$, $A(2) = 0$, and $A(2\sqrt{3}/3) \geq 0$, the maximum possible area is $32\sqrt{3}/9$.

19. Let x be the length of the edge of each of the twelve small squares. Then each of the three cross-shaped pieces will form boxes with base length $1 - 2x$ and height x, so each of the three will have volume $x(1 - 2x)^2$. Both of the two cubical boxes will have edge x and thus volume x^3. So the total volume of all five boxes will be

$$V(x) = 3x(1 - 2x)^2 + 2x^3 = 14x^3 - 12x^2 + 3x, \qquad 0 \leq x \leq \frac{1}{2}.$$

Now $V'(x) = 42x^2 - 24x + 3$; $V'(x) = 0$ when $14x^2 - 8x - 1 = 0$. The quadratic formula gives the two solutions $\dfrac{4 \pm \sqrt{2}}{14}$. These are approximately 0.3867 and 0.1847, and both lie in the domain of V. Now $V(0) = 0$, $V(0.1847) \approx 0.2329$, $V(0.3867) \approx 0.1752$, and $V(0.5) = 0.25$. Therefore, to maximize V, one must cut each of the three large squares into four smaller squares of side length $\frac{1}{2}$ each and form the resulting twelve squares into two cubes. At maximum volume there will be only two boxes, not five.

22. Let r be the radius of the circle and x the edge of the square. We are to maximize total area $A = \pi r^2 + x^2$ given the side condition $2\pi r + 4x = 100$. From the last equation we infer that

$$x = \frac{100 - 2\pi r}{4} = \frac{50 - \pi r}{2}.$$

So

$$A = A(r) = \pi r^2 + \frac{1}{4}(50 - \pi r)^2 = (\pi + \frac{1}{4}\pi^2)r - 25\pi r + 625$$

for $0 \leq r \leq 50/\pi$ (because $x \geq 0$). Now

$$A'(r) = 2(\pi + \frac{1}{4}\pi^2)r - 25\pi;$$

$$A'(r) = 0 \text{ when } r = \frac{25}{2 + \dfrac{\pi}{2}} = \frac{50}{\pi + 4};$$

that is, when $r \approx 7$. Finally, $A(0) = 625$,

$$A\left(\frac{50}{\pi}\right) \approx 795.77 \quad \text{and} \quad A\left(\frac{50}{\pi + 4}\right) \approx 350.06$$

Results: For minimum area, use a circle of radius $50/(\pi + 4) \approx 7.00124$ (cm) and a square of edge length $100/(\pi + 4)$ (cm). For maximum area, bend all the wire into a circle of radius $50/\pi$.

25. Let the dimensions of the box be x by x by y. We are to maximize $V = x^2 y$ subject to some conditions on x and y. According to the poster on the wall of the Bogart, Georgia, Post Office, the *length* of the box is the larger of x and y, and the *girth* is measured around the box in a plane perpendicular to its length.

Case 1: $x < y$. Then the length is y, the girth is $4x$, and the mailing constraint is $4x + y \leq 100$. It is clear that we take $4x + y = 100$ to maximize V, so that

$$V = V(x) = x^2(100 - 4x) = 100x^2 - 4x^3, \quad 0 \leq x \leq 25.$$

Then $V'(x) = 4x(50 - 3x)$; $V'(x) = 0$ for $x = 0$ and for $x = 50/3$. But $V(0) = 0$, $V(25) = 0$, and $V(50/3) = 250,000/27 \approx 9259$ (in.^{3}). The latter is the maximum in Case 1.

Case 2: $x \geq y$. Then the length is x and the girth is $2x + y$, though you may get some argument from a postal worker who may insist that it's $4x$. So $3x + 2y = 100$, and thus

$$V = V(x) = x^2\left(\frac{100 - 3x}{2}\right) = 50x^2 - \frac{3}{2}x^3, \quad 0 \leq x \leq 100/3.$$

Then $V'(x) = 100x - \frac{9}{2}x^2$; $V'(x) = 0$ when $x = 0$ and when $x = 200/9$. But $V(0) = 0$, $V(100/3) = 0$, and $V(200/9) = 2,000,000/243 \approx 8230$ (in.^{3}).

Case 3: You lose the argument in Case 2. Then the box has length x and girth $4x$, so $5x = 100$; thus $x = 20$. To maximize the total volume, no calculus is needed—let $y = x$. Then the box of maximum volume will have volume $20^3 = 8000$ (in.^{3}).

Answer: The maximum is $\dfrac{250,000}{27}$ in.^{3}.

28. Let x denote the number of workers hired. Each worker will pick $900/x$ bushels; each worker will spend $180/x$ hours picking beans. The supervisor cost will be $1800/x$ dollars, and the cost per worker will be $8 + 900/x$ dollars. Thus the total cost will be

$$C(x) = 8x + 900 + \frac{1800}{x}, \quad 1 \leq x.$$

It is clear that large values of x make $C(x)$ large, so the global minimum of $C(x)$ occurs either at $x = 1$ or where $C'(x) = 0$. Assume for the moment that x can take on all real number values in $[1, \infty)$, not merely integral values, so that C' is defined. Then

$$C'(x) = 8 - \frac{1800}{x^2}; \quad C'(x) = 0 \text{ when } x^2 = 225.$$

Thus $C'(15) = 0$. Now $C(1) = 2708$ and $C(15) = 1140$, so fifteen workers should be hired; the cost to pick each bushel will be approximately $1.27.

34. Let the circle have equation $x^2 + y^2 = 1$ and let (x, y) denote the coordinates of the upper right-hand vertex of the trapezoid (Fig. 3.6.25). Then the area A of the trapezoid is the product of its altitude y and the average of the lengths of its two bases, so

$$A = \frac{1}{2}y(2x + 2) \text{ where } y^2 = 1 - x^2.$$

A positive quantity is maximized when its square is maximized, so we maximize instead

$$f(x) = A^2 = (x + 1)^2(1 - x^2)$$
$$= 1 + 2x - 2x^3 - x^4, \quad 0 \le x \le 1.$$

Because $f(0) = 0 = f(1)$, f is maximized when $f'(x) = 0$:

$$0 = 2 - 6x^2 - 4x^3 = 2(1 + x)^2(1 - 2x).$$

But the only solution of $f'(x) = 0$ in the domain of f is $x = 1/2$. Finally, $f(1/2) = 27/16$, so the maximum possible area of the trapezoid is $\frac{3}{4}\sqrt{3}$. This is just over 41% of the area of the circle, so the answer meets the test of plausibility.

37. We are to maximize volume $V = \frac{1}{3}\pi r^2 h$ given $r^2 + h^2 = 100$. The latter relation enables us to write

$$V = V(h) = \frac{1}{3}\pi(100 - h^2)h = \frac{1}{3}\pi(100h - h^3), \quad 0 \le h \le 10.$$

Now $V'(h) = \frac{1}{3}\pi(100 - 3h^2)$, so $V'(h) = 0$ when $3h^2 = 100$, thus when $h = \frac{10}{3}\sqrt{3}$. But $V(h) = 0$ at the endpoints of its domain, so the latter value of h maximizes V, and its maximum value is $\frac{2000}{27}\pi\sqrt{3}$.

40. If the base of the L has length x, then the vertical part has length $60 - x$. Place the L with its corner at the origin in the xy-plane, its base on the nonnegative x-axis, and the vertical part on the nonnegative y-axis. The two ends of the L have coordinates $(0, 60 - x)$ and $(x, 0)$, so they are at distance

$$d = d(x) = \sqrt{x^2 + (60 - x)^2}, \quad 0 \le x \le 60.$$

A positive quantity is minimized when its square is minimal, so we minimize

$$f(x) = d^2 = x^2 + (60 - x)^2, \quad 0 \le x \le 60.$$

Then $f'(x) = 2x - 2(60 - x) = 4x - 120$; $f'(x) = 0$ when $x = 30$. Now $f(0) = f(60) = 3600$, whereas $f(30) = 1800$. So $x = 30$ minimizes $f(x)$ and thus $d(x)$. The minimum possible distance between the two ends of the wire is therefore $d(30) = 30\sqrt{2}$.

43. Examine the plank on the right on Fig. 3.6.10. Let its height be $2y$ and its width (in the x-direction) be z. The total area of the four small rectangles in the figure is then $A = 4 \cdot z \cdot 2y = 8yz$. The circle has radius 1, and by Problem 35 the large inscribed square has dimensions $\sqrt{2}$ by $\sqrt{2}$. Thus

$$\left(\frac{1}{2}\sqrt{2} + z\right)^2 + y^2 = 1;$$

This implies that

$$y = \sqrt{\frac{1}{2} - z\sqrt{2} - z^2}.$$

Therefore

$$A(z) = 8z\sqrt{\frac{1}{2} - z\sqrt{2} - z^2}, \qquad 0 \le z \le 1 - \frac{1}{2}\sqrt{2}.$$

Now $A = 0$ at each endpoint of its domain, and

$$A'(z) = \frac{4\sqrt{2}(1 - 3z\sqrt{z} - 4z^2)}{\sqrt{1 - 2z\sqrt{2} - 2z^2}}.$$

So $A'(z) = 0$ when $z = \dfrac{-3\sqrt{2} \pm \sqrt{34}}{8}$; we discard the negative solution, and find that when $A(z)$ is maximized,

$$z = \frac{-3\sqrt{2} + \sqrt{34}}{8} \approx 0.198539,$$

$$2y = \frac{\sqrt{7 - \sqrt{17}}}{2} \approx 0.848071, \text{ and}$$

$$A(z) = \frac{\sqrt{142 + 34\sqrt{17}}}{2} \approx 0.673500.$$

The four small planks use just under 59% of the wood that remains after the large plank is cut, a very efficient use of what might be scrap lumber.

46. Set up a coordinate system in which the factory is located at the origin and the power station at (L, W) in the xy-plane—$L = 4500$, $W = 2000$. Part of the path of the power cable will be straight along the river bank and part will be a diagonal running under water. It makes no difference whether the straight part is adjacent to the factory or to the power station, so we assume the former. Thus we suppose that the power cable runs straight from $(0,0)$ to $(x,0)$, then straight from $(x,0)$ to (L, W), where $0 \le x \le L$. Let y be the length of the diagonal stretch of the cable. Then by the Pythagorean theorem,

$$W^2 + (L - x)^2 = y^2, \text{ so } y = \sqrt{W^2 + (L - x)^2}.$$

The cost C of the cable is $C = kx + 3ky$ where k is the cost per unit distance of over-the-ground cable. Therefore the total cost of the cable is

$$C(x) = kx + 3k\sqrt{W^2 + (L - x)^2}, \qquad 0 \le x \le L.$$

It will not change the solution if we assume that $k = 1$, and in this case we have

$$C'(x) = 1 - \frac{3(L - x)}{\sqrt{W^2 + (L - x)^2}}.$$

Next, $C'(x) = 0$ when $x^2 + (L - x)^2 = 9(L - x)^2$, and this leads to the solution

$$x = L - \frac{1}{4}W\sqrt{2} \text{ and } y = \frac{3}{4}W\sqrt{2}.$$

It is not difficult to verify that the latter value of x yields a value of C smaller than either $C(0)$ or $C(L)$. Answer: Lay the cable $x = 4500 - 500\sqrt{2} \approx 3793$ meters along the bank and $y = 1500\sqrt{2} \approx 2121$ meters diagonally across the river.

49. We are to minimize total cost

$$C = c_1\sqrt{a^2 + x^2} + c_2\sqrt{(L - x)^2 + b^2}.$$

$$C'(x) = \frac{c_1 x}{\sqrt{a^2 + x^2}} - \frac{c_2(L-x)}{\sqrt{(L-x)^2 + b^2}};$$

$$C'(x) = 0 \text{ when } \frac{c_1 x}{\sqrt{a^2 + x^2}} = \frac{c_2(L-x)}{\sqrt{(L-x)^2 + b^2}}.$$

The result in Part (a) is equivalent to the last equation. For Part (b), assume that $a = b = c_1 = 1$, $c_2 = 2$, and $L = 4$. Then we obtain

$$\frac{x}{\sqrt{1+x^2}} = \frac{2(4-x)}{\sqrt{(4-x)^2 + 1}};$$

$$\frac{x^2}{1+x^2} = \frac{4(16 - 8x + x^2)}{16 - 8x + x^2 + 1};$$

$$x^2(17 - 8x + x^2) = (4 + 4x^2)(16 - 8x + x^2);$$

$$17x^2 - 8x^3 + x^4 = 64 - 32x + 68x^2 - 32x^3 + 4x^4.$$

Therefore we wish to solve $f(x) = 0$ where

$$f(x) = 3x^4 - 24x^4 + 51x^2 - 32x + 64.$$

Now $f(0) = 64$, $f(1) = 62$, $f(2) = 60$, $f(3) = 22$, and $f(4) = -16$. Because $f(3) > 0 > f(4)$, we interpolate to estimate the zero of $f(x)$ between 3 and 4; it turns out that interpolation gives $x \approx 3.58$. Subsequent interpolation yields the more accurate estimate $x \approx 3.45$. (The equation $f(x) = 0$ has exactly two solutions, $x \approx 3.452462314$ and $x \approx 4.559682567$.)

52. Let the horizontal piece of wood have length $2x$ and the vertical piece have length $y + z$ where y is the length of the part above the horizontal piece and z the length of the part below it. Then

$$y = \sqrt{4 - x^2} \text{ and } z = \sqrt{16 - x^2}.$$

Also

$$y + z = \frac{x^2}{y} + \frac{x^2}{z}.$$

Multiply each side of the last equation by yz to obtain

$$y^2 z + yz^2 = x^2 z + x^2 y,$$

so that

$$yz(y + z) = x^2(y + z);$$

$$x^2 = yz;$$

$$x^4 = y^2 z^2 = (4 - x^2)(16 - x^2);$$

$$x^4 = 64 - 20x^2 + x^4;$$

$$20x^2 = 64;$$

$$x = \frac{4}{5}\sqrt{5}, \ y = \frac{2}{5}\sqrt{5}, \ z = \frac{8}{5}\sqrt{5}.$$

Therefore $L_1 = \frac{8}{5}\sqrt{5} \approx 3.5777$ and $L_2 = 2\sqrt{5} \approx 4.47214$ for maximum area.

Section 3.7

1. $f'(x) = 6 \sin x \cos x$

4. $f'(x) = \dfrac{1}{2\sqrt{x}} \sin x + \sqrt{x} \cos x$

7. $f'(x) = \cos^3 x - 2\sin^2 x \cos x$

10. $g'(t) = 6(2 - \cos^2 t)^2 \sin t \cos t$

13. $f'(x) = 2\sin x + 2x \cos x - 6x \cos x + 3x^2 \sin x$

16. $f'(x) = 7\cos 5x \cos 7x - 5\sin 5x \sin 7x$

19. $g'(t) = -\frac{5}{2}(\cos 3t + \cos 5t)^{3/2}(3\sin 3t + 5\sin 5t)$

22. $\dfrac{dy}{dx} = \dfrac{-2x\sin 2x - \cos 2x}{x^2}$

25. $\dfrac{dy}{dx} = 2\cos 2x \cos 3x - 3\sin 2x \sin 3x$

28. $\dfrac{dy}{dx} = -\dfrac{\sin\sqrt{x}}{4\sqrt{x}\sqrt{\cos\sqrt{x}}}$

31. $\dfrac{dy}{dx} = \dfrac{\cos 2\sqrt{x}}{\sqrt{x}}$

34. $\dfrac{dy}{dx} = 2x\cos\dfrac{1}{x} + \sin\dfrac{1}{x}$

37. $\dfrac{dy}{dx} = \frac{1}{2}x^{-1/2}(x - \cos x)^3 + 3x^{1/2}(x - \cos x)^2(1 + \sin x)$

40. $\dfrac{dy}{dx} = \dfrac{(\cos x)\cos(1 + \sqrt{\sin x})}{2\sqrt{\sin x}}$

43. $\dfrac{dy}{dx} = 7\sec^2 x \tan^6 x$

46. $\dfrac{dy}{dx} = \dfrac{5x^5 \sec x^5 \tan x^5 - \sec x^5}{x^2}$

49. $\dfrac{dy}{dx} = \dfrac{2\cot\dfrac{1}{x^2}\csc\dfrac{1}{x^2}}{x^3}$

52. $\dfrac{dy}{dx} \equiv 0$

55. $\dfrac{dy}{dx} = [\sec(\sin x)\,\tan(\sin x)]\cos x$

58. $\dfrac{dy}{dx} = \dfrac{(1 + \tan x)\sec x \tan x - \sec^3 x}{(1 + \tan x)^2} = \dfrac{\sec x \tan x - \sec x}{(1 + \tan x)^2}$

61.   For example, if $f(x) = \sec x = \dfrac{1}{\cos x}$, then

$$f'(x) = -\frac{-\sin x}{\cos^2 x} = \frac{1}{\cos x} \cdot \frac{\sin x}{\cos x} = \sec x \tan x.$$

64.   Let h be the altitude of the balloon (in feet) at time t (in seconds) and let θ be its angle of elevation with respect to the observer. From the obvious figure, $h = 300\tan\theta$, so

$$\frac{dh}{dt} = (300\sec^2\theta)\frac{d\theta}{dt}.$$

When $\theta = \pi/4$ and $\dfrac{d\theta}{dt} = \pi/180$, we have

$$\frac{dh}{dt} = 300 \cdot 2 \cdot \frac{\pi}{180} = \frac{10\pi}{3} \approx 10.47 \text{ ft/s}$$

as the rate of the balloon's ascent then.

67.   As in the figure on the next page, let θ be the angle of elevation of the observer's line of sight. Then

$$\tan\theta = \frac{20000}{x},$$

so that $x = 20000 \cot \theta$. Thus

$$\frac{dx}{dt} = (-20000 \csc^2 \theta)\frac{d\theta}{dt}.$$

When $\theta = 60°$, we are given $\dfrac{d\theta}{dt} = 0.5°/\text{s}$; that is,

$\dfrac{d\theta}{dt} = \pi/360$ radians per second when $\theta = \pi/3$. We evaluate dx/dt at this time with these values to obtain

$$\frac{dx}{dt} = (-20000)\frac{1}{\sin^2\left(\dfrac{\pi}{3}\right)} \cdot \frac{\pi}{360} = -\frac{2000\pi}{27},$$

approximately -232.71 ft/s. Answer: About 158.67 mi/h.

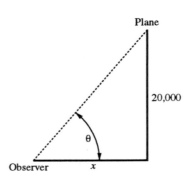

Plane

20,000

θ

Observer x

70. In the situation described in the problem, we have $D = 20 \sec \theta$. The illumination of the walkway is then

$$I = I(\theta) = \frac{k}{400} \sin \theta \cos^2 \theta, \qquad 0 \le \theta \le \pi/2.$$

$$\frac{dI}{d\theta} = \frac{k \cos \theta}{400}(\cos^2 \theta - 2\sin^2 \theta);$$

$dI/d\theta = 0$ when $\theta = \pi/2$ and when $\cos^2 \theta = 2\sin^2 \theta$. The solution θ in the domain of I of the latter equation has the property that $\sin \theta = \sqrt{3}/3$ and $\cos \theta = \sqrt{6}/3$. But $I(0) = 0$ and $I(\theta) \to 0$ as $\theta \to (\pi/2)^-$, so the optimal height of the lamp post occurs when $\sin \theta = \sqrt{3}/3$. This implies that the optimal height is $10\sqrt{2} \approx 14.14$ m.

73. Set up coordinates so the diameter is on the x-axis and the equation of the circle is $x^2 + y^2 = 1$; let (x, y) denote the northwest corner of the trapezoid. The chord from $(1, 0)$ to (x, y) forms a right triangle with hypotenuse 2, side z opposite angle θ, and side w; moreover, $z = 2 \sin \theta$ and $w = 2 \cos \theta$. It follows that

$$y = w \sin \theta = 2 \sin \theta \cos \theta \text{ and}$$
$$-x = 1 - w \cos \theta = -\cos 2\theta.$$

Now

$$A = y(1 - x) = (2 \sin \theta \cos \theta)(1 - \cos 2\theta)$$
$$= 4 \sin \theta \cos \theta \sin^2 \theta,$$

and therefore

$$A = A(\theta) = 4 \sin^3 \theta \cos \theta, \qquad \frac{\pi}{4} \le \theta \le \frac{\pi}{2}.$$

$$A'(\theta) = 12 \sin^2 \theta \cos^2 \theta - 4 \sin^4 \theta$$
$$= (4 \sin^2 \theta)(3 \cos^2 \theta - \sin^2 \theta).$$

To solve $A'(\theta) = 0$, we note that $\sin \theta \ne 0$, so we must have $3 \cos^2 \theta = \sin^2 \theta$; that is $\tan^2 \theta = 3$. It follows that $\theta = \pi/3$. The value of A here exceeds its value at the endpoints, so we have found the maximum value of the area—it is $3\sqrt{3}/4$.

76. The length of the forest path is $2 \csc \theta$. So the length of the part of the trip along the road is $3 - 2 \csc \theta \cos \theta$. Thus the total time for the trip is given by

$$T = T(\theta) = \frac{2}{3 \sin \theta} + \frac{3 - \dfrac{2 \cos \theta}{\sin \theta}}{8}.$$

Note that the range of values of θ is determined by the condition

$$\frac{3\sqrt{13}}{13} \geq \cos\theta \geq 0.$$

After simplifications, we find that

$$T'(\theta) = \frac{3 - 8\cos\theta}{12\sin^2\theta}.$$

Now $T'(\theta) = 0$ when $\cos\theta = 3/8$; that is, when θ is approximately $67°58'32"$. For this value of θ, we find that $\sin\theta = \frac{1}{8}\sqrt{55}$. There's no problem in verifying that we have found the minimum. Answer: The distance to walk down the road is

$$\left(3 - 2\frac{\cos\theta}{\sin\theta}\right)\bigg|_{\sin\theta = \frac{\sqrt{55}}{8}} = 3 - \frac{6\sqrt{55}}{55} \approx 2.19096 \text{ (km)}.$$

Section 3.8

1. $2x - 2y\dfrac{dy}{dx} = 0$: $\quad \dfrac{dy}{dx} = \dfrac{x}{y}$.

4. $3x^2 + 3y^2\dfrac{dy}{dx} = 0$: $\quad \dfrac{dy}{dx} = -\dfrac{x^2}{y^2}$.

7. $\frac{2}{3}x^{-1/3} + \frac{2}{3}y^{-1/3}\dfrac{dy}{dx} = 0$: $\quad \dfrac{dy}{dx} = -\left(\dfrac{y}{x}\right)^{1/3}$

10. $2(x^2 + y^2)\left(2x + 2y\dfrac{dy}{dx}\right) = 4x\dfrac{dy}{dx} + 4y$:

$$[(2x^2 + 2y^2)(2y) - 4x]\frac{dy}{dx} = 4y - 4x(x^2 + y^2);$$

$$\frac{dy}{dx} = \frac{y - x^3 - xy^2}{x^2y + y^3 - x}.$$

13. $x^2\dfrac{dy}{dx} + 2xy = 1$, so $\dfrac{dy}{dx} = \dfrac{1 - 2xy}{x^2}$. At $(2,1)$ the tangent has slope $-\frac{3}{4}$ and thus equation $3x + 4y = 10$.

16. $-\dfrac{1}{(x+1)^2} - \dfrac{1}{(y+1)^2} \cdot \dfrac{dy}{dx} = 0$, so $\dfrac{dy}{dx} = -\dfrac{(y+1)^2}{(x+1)^2}$. At $(1,1)$ the tangent line has slope -1 and thus equation $y - 1 = -(x - 1)$.

19. $-3x^{-4} - 3y^{-4}\dfrac{dy}{dx} = 0$: $\quad \dfrac{dy}{dx} = -\dfrac{y^4}{x^4}$. At $(1,1)$ the tangent has slope -1, and thus equation $y - 1 = -(x - 1)$.

22. Here we have $y^5 + 5xy^4\dfrac{dy}{dx} + 5x^4y + x^5\dfrac{dy}{dx} = 0$:

$$\frac{dy}{dx} = -\frac{y^5 + 5x^4y}{x^5 + 5xy^4}.$$

A fraction can be zero only when its numerator is zero, so $\dfrac{dy}{dx} = 0$ only if

$$y^5 + 5x^4y = y(y^4 + 5x^4) = 0.$$

This can hold only if $y = 0$ or if $y^4 + 5x^4 = 0$, and in either case $y = 0$. But there are no points on the graph of $xy^5 + x^5y = 1$ corresponding to $y = 0$. Therefore the graph has no horizontal tangents.

25. Here $\dfrac{dy}{dx} = \dfrac{2 - x}{y - 2}$, so horizontal tangents can occur only if $x = 2$ and $y \neq 2$. When $x = 2$, the given equation yields $y^2 - 4y - 4 = 0$, so that $y = 2 \pm \sqrt{8}$. Thus there are two points at which the tangent line is horizontal.

28. Here, $\dfrac{dy}{dx} = \dfrac{y - 2x}{2y - x}$ and $\dfrac{dx}{dy} = \dfrac{2y - x}{y - 2x}$. For horizontal tangents, $y = 2x$, and from the original equation $x^2 - xy + y^2 = 9$; it follows upon substitution of $2x$ for y that $x^2 = 3$. So the tangent line is horizontal at $(\sqrt{3}, 2\sqrt{3})$ and at $(-\sqrt{3}, -2\sqrt{3})$. Where there are vertical tangents we must have $x = 2y$, and (as above) it turns out that $y^2 = 3$; there are vertical tangents at $(2\sqrt{3}, \sqrt{3})$ and at $(-2\sqrt{3}, -\sqrt{3})$.

31. Suppose that the pile has height $h = h(t)$ at time t (seconds) and radius $r = r(t)$ then. We are given $h = 2r$ and we know that the volume of the pile at time t is

$$V = V(t) = \frac{\pi}{3}r^2 h = \frac{2}{3}\pi r^3. \quad \text{Now} \quad \frac{dV}{dt} = \frac{dV}{dr} \cdot \frac{dr}{dt}, \text{ so } 10 = 2\pi r^2 \frac{dr}{dt}.$$

When $h = 5$, $r = 2.5$; at that time $\dfrac{dr}{dt} = \dfrac{10}{2\pi(2.5)^2} = \dfrac{4}{5\pi} \approx 0.25645$ (ft/s).

34. Let x be the distance from the ostrich to the street light and u the distance from the base of the light pole to the tip of the ostrich's shadow. Draw a figure and so label it; by similar triangles you find that $\dfrac{u}{10} = \dfrac{u - x}{5}$, and it follows that $u = 2x$. We are to find du/dt and $D_t(u - x) = du/dt - dx/dt$. But $u = 2x$, so

$$\frac{du}{dt} = 2\frac{dx}{dt} = (2)(-4) = -8; \qquad \frac{du}{dt} - \frac{dx}{dt} = -8 - (-4) = -4.$$

Answers: (a) $+8$ ft/s; (b) $+4$ ft/s.

37. Let r denote the radius of the balloon and V its volume at time t (in seconds). Then

$$V = \frac{4}{3}\pi r^3, \text{ so } \frac{dV}{dt} = 4\pi r^2 \frac{dr}{dt}.$$

We are to find dr/dt when $r = 10$, and we are given the information that $dV/dt = 100\pi$. Therefore

$$100\pi = 4\pi(10)^2 \frac{dr}{dt}\bigg|_{r=10},$$

and so at the time in question the radius is increasing at the rate of $dr/dt = \frac{1}{4} = 0.25$ (cm/s).

40. Locate the observer at the origin and the balloon in the first quadrant at $(300, y)$, where $y = y(t)$ is the balloon's altitude at time t. Let θ be the angle of elevation of the balloon (in radians) from the observer's point of view. Then $\tan\theta = y/300$. We are given $d\theta/dt = \pi/180$ rad/s. Hence we are to find dy/dt when $\theta = \pi/4$. But $y = 300\tan\theta$, so

$$\frac{dy}{dt} = (300\sec^2\theta)\frac{d\theta}{dt}.$$

Substitution of the given values of θ and $d\theta/dt$ yields the answer

$$\frac{dy}{dt}\bigg|_{\theta=45°} = 300 \cdot 2 \cdot \frac{\pi}{180} = \frac{10\pi}{3} \approx 10.472 \text{ (ft/s)}.$$

43. We use $a = 10$ in the formula given in Problem 42. Then

$$V = \frac{1}{3}\pi y^2 (30 - y).$$

Hence $(-100)(0.1337) = \dfrac{dV}{dt} = \pi(20y - y^2)\dfrac{dy}{dt}$. Thus $\dfrac{dy}{dt} = -\dfrac{13 \cdot 37}{\pi y(20 - y)}$. Substitution of $y = 7$ and $y = 3$ now yields the two answers.

46. Let x be the length of the base of the rectangle and y its height. We are given $dx/dt = +4$ and $dy/dt = -3$, with units in centimeters and seconds. The area of the rectangle is $A = xy$, so

$$\frac{dA}{dt} = x\frac{dy}{dt} + y\frac{dx}{dt} = -3x + 4y.$$

Therefore when $x = 20$ and $y = 12$, we have $dA/dt = -12$, so the area of the rectangle is decreasing at the rate of 12 cm^2/s then.

49. Locate the radar station at the origin and the rocket at $(4, y)$ in the first quadrant at time t, with y in miles and t in hours. The distance z between the station and the rocket satisfies the equation $y^2 + 16 = z^2$, so $2y\dfrac{dy}{dt} = 2z\dfrac{dz}{dt}$. When $z = 5$, we have $y = 3$, and because $dz/dt = 3600$ it follows that $dy/dt = 6000$ mi/h.

52. Let x be the distance between the Pinta and the island at time t and y the distance between the Niña and the island then. We know that $x^2 + y^2 = z^2$ where $z = z(t)$ is the distance between the two ships, so

$$2z\frac{dz}{dt} = 2x\frac{dx}{dt} + 2y\frac{dy}{dt}.$$

When $x = 30$ and $y = 40$, $z = 50$. It follows from the last equation that $dz/dt = -25$ then. Answer: They are drawing closer at 25 mi/h then.

55. Let x be the radius of the water surface at time t and y the height of the water remaining at time t. If Q is the amount of water remaining in the tank at time t, then (because the water forms a cone) $Q = Q(t) = \frac{1}{3}\pi x^2 y$. But by similar triangles, $\dfrac{x}{y} = \dfrac{3}{5}$, so $x = \dfrac{3y}{5}$. So

$$Q(t) = \frac{1}{3}\pi\frac{9}{25}y^3 = \frac{3}{25}\pi y^3.$$

We are given $dQ/dt = -2$ when $y = 3$. This implies that when $y = 3$, $-2 = \dfrac{dQ}{dt} = \dfrac{9}{25}\pi y^2\dfrac{dy}{dt}$. So at the time in question,

$$\left.\frac{dy}{dt}\right|_{y=3} = -\frac{50}{81\pi} \approx -0.1965 \text{ (ft/s)}.$$

58. Let x denote the distance between the ship and A, y the distance between the ship and B, h the perpendicular distance from the position of the ship to the line AB, u the distance from A to the foot of this perpendicular, and v the distance from B to the foot of the perpendicular. At the time in question, we know that $x = 10.4$, $dx/dt = 19.2$, $y = 5$, and $dy/dt = -0.6$. From the right triangles involved, we see that $u^2 + h^2 = x^2$ and $(12.6 - u)^2 + h^2 = y^2$. Therefore

$$x^2 - u^2 = y^2 - (12.6 - u)^2. \tag{$*$}$$

We take $x = 10.4$ and $y = 5$ in Eq. ($*$); it follows that $u = 9.6$ and that $v = 12.6 - u = 3$. From Eq. ($*$), we know that

$$x\frac{dx}{dt} - u\frac{du}{dt} = y\frac{dy}{dt} + (12.6 - u)\frac{du}{dt},$$

so

$$\frac{du}{dt} = \frac{1}{12.6}\left(x\frac{dx}{dt} - y\frac{dy}{dt}\right).$$

From the data given, $du/dt \approx 16.0857$. Also, because $h = \sqrt{x^2 - u^2}$, $h = 4$ when $x = 10.4$ and $y = 9.6$. Moreover, $h\dfrac{dh}{dt} = x\dfrac{dx}{dt} - u\dfrac{du}{dt}$, and therefore

$$\left.4\frac{dh}{dt}\right|_{h=4} \approx (10.4)(19.2) - (9.6)(16.0857) \approx 11.3143.$$

Finally, $\dfrac{dh/dt}{du/dt} \approx 0.7034$, so the ship is sailing a course about $35°7'$ north or south of east at a speed of $\sqrt{(du/dt)^2 + (dh/dt)^2} \approx 19.67$ mi/h. It is located 9.6 miles east and 4 miles north or south of A, or 10.4 miles from A at a bearing of either $67°22'\,48"$ or $112°37'\,12"$.

61. Let r denote the radius of the conical pile and h its height. Then $r = 3h$, and so the volume of the pile is $V = 3\pi h^3$. We are given $dV/dt = 120\pi$, so

$$120\pi = \frac{dV}{dt} = 9\pi h^2 \frac{dh}{dt}.$$

When $h = 20$, we therefore have $120\pi = 3600\pi \frac{dh}{dt}$, and thus when the pile is 20 feet high, its altitude is increasing at $1/30$ of a foot per second—2 ft/min.

64. Set up a coordinate system in which the officer is at the origin, the van is moving in the positive direction along the line $y = 200$ (so units on the coordinate axes are in feet). When the van is at position $(x, 200)$, the distance from the officer to the van is z, where $x^2 + 200^2 = z^2$, so that $x\frac{dx}{dt} = z\frac{dz}{dt}$. When the van reaches the call box, $x = 200$, $z = 200\sqrt{2}$, and $dz/dt = 66$. It follows that

$$\left.\frac{dx}{dt}\right|_{x=200} = 66\sqrt{2},$$

which translates to about 63.6 mi/h.

Section 3.9

Note: In this section your results may differ from the answers in the last one or two decimal places because of differences in calculators or in methods of solving the equations.

Note: In Problems 1 through 20, we obtained our initial estimate x_0 of the solution by linear interpolation: Write the equation of the straight line that joins $(a, f(a))$ with $(b, f(b))$ and let x_0 be the x-coordinate of the point where this line crosses the x-axis.

1. $x_0 = 2.2$; we use $f(x) = x^2 - 5$.
   Then $x_1 = 2.236363636$, $x_2 = 2.236067997$, and $x_3 = x_4 = 2.236067978$.

4. Let $f(x) = x^{3/2} - 10$. Then $x_0 = 4.628863603$. From the iterative formula

$$x \longleftarrow x - \frac{x^{3/2} - 10}{\frac{3}{2}x^{1/2}}$$

   we obtain $x_1 = 4.641597575$, $x_2 = 4.641588834 = x_3$.

7. $x_0 = -0.5$;
   $x_1 = -0.8108695652$, $x_2 = -0.7449619516$, $x_3 = -0.7402438226$, $x_4 = -0.7402217821 = x_5$

10. Let $f(x) = x^2 - \sin x$. Then $f'(x) = 2x - \cos x$. The interpolation formula at the beginning of this solution section yields $x_0 = 0.7956861008$, and the iterative formula

$$x \longleftarrow x - \frac{x^2 - \sin x}{2x - \cos x}$$

   (with calculator in *radian* mode) yields these results:
   $x_1 = 0.8867915207$, $x_2 = 0.8768492470$, $x_3 = 0.8767262342$, and $x_4 = 0.8767262154 = x_5$.

13. With $x_0 = 2.188405797$ and the iterative formula

$$x \longleftarrow x - \frac{x^4(x+1) - 100}{x^3(5x+4)},$$

   we obtain $x_1 = 2.360000254$, $x_2 = 2.339638357$, $x_3 = 2.339301099$, and $x_4 = 2.339301008 = x_5$.

16. Because $7\pi/2 \approx 10.9956$ and $4\pi \approx 12.5663$ are the nearest discontinuities of $f(x) = x - \tan x$, this function has the intermediate value property on the interval $[11, 12]$. Because $f(11) \approx -214.95$ and

$f(12) \approx 11.364$, the equation $f(x) = 0$ has a solution in $[11, 12]$. We obtain $x_0 = 11.94978618$ by interpolation, and the iteration

$$x \longleftarrow x - \frac{x + \tan x}{1 + \sec^2 x}$$

of Newton's method yields the successive approximations

$$x_1 = 7.457596948, \quad x_2 = 6.180210620, \quad x_3 = 3.157913273, \quad x_4 = 1.571006986;$$

after many more iterations we arrive at the answer 2.028757838 of Problem 15. The difficulty is caused by the fact that $f(x)$ is generally a very large number, so the iteration of Newton's method tends to alter the value of x excessively. A little experimentation yields the fact that $f(11.08) \approx -0.736577$ and $f(11.09) \approx 0.531158$. We begin anew on the better interval $[11.08, 11.09]$ and obtain $x_0 = 11.08581018$, $x_1 = 11.08553759$, $x_2 = 11.08553841$, and $x_3 = x_2$.

19. $x_0 = 1.538461538$, $\qquad x_1 = 1.932798309$, $\qquad x_2 = 1.819962709$,
    $x_3 = 1.802569044$, $\qquad x_4 = 1.802191039$, $\qquad x_5 = 1.802190864$,
    $x_6 = 1.802190864 = x_5$.
    The convergence is slow because $|f'(x)|$ is large when x is near 1.8.

22. (b) With $x_0 = 1.5$ we obtain the successive approximations 1.610123, 1.586600, 1.584901, 1.584893, and 1.584893.

25. Using the first formula, we obtain $x_0 = 0.5$, $x_1 = -1$, $x_2 = 2$, $x_3 = 2.75$, $x_4 = 2.867768595$, ..., $x_{12} = 2.879385242 = x_{13}$. Wrong root! At least the method converged. With the second formula, we find $x_0 = 0.5$, error message on x_1. That's because our pocket computer won't compute the cube root of a negative number. With the alternative formula $x = \sqrt[3]{3x^2 - 1}$, we obtain $x_0 = 0.5$, $x_1 = 0.629960525$, $x_2 = 0.5754446861$, $x_3 = 0.1874852347$, $x_4 = 0.9635358111$, $x_5 = 1.213098128$, ..., $x_{62} = 2.879385240 = x_{63}$. Wrong root again. Actually, we're lucky to get a correct root in any case, because using absolute values in the formula changes it so much that we might be solving the wrong equation!

28. 0.8241323123 and -0.8241323123

34. The graphs of $y = x$ and $y = \tan x$ show that the smallest positive solution of $f(x) = x - \tan x = 0$ is between π and $3\pi/2$. With initial guess $x = 4.5$ we obtain 4.493613903, 4.493409655, 4.493409458, and 4.493409458.

37. With $x_0 = 0.25$, Newton's method yields the sequence 0.2259259259, 0.2260737086, 0.2260737138, 0.2260717138. To four places, $x = 0.2261$.

Chapter 3 Miscellaneous

1. $\dfrac{dy}{dx} = 2x - \dfrac{6}{x^3}$

4. $\dfrac{dy}{dx} = \frac{5}{2}(x^2 + 4x)^{3/2}(2x + 4) = 5(x + 2)(x^2 + 4x)^{3/2}$

7. $\dfrac{dy}{dx} = 4(3x - \frac{1}{2}x^{-2})^3(3 + x^{-3})$

10. $y = (5x^6)^{-1/2}$: $\quad \dfrac{dy}{dx} = -\frac{1}{2}(5x^6)^{-3/2}(30x^5) = -\dfrac{3}{x\sqrt{5x^6}} = -\dfrac{3\sqrt{5}}{5x^4}$

13. $\dfrac{dy}{dx} = \dfrac{dy}{du} \cdot \dfrac{du}{dx} = \dfrac{-2u}{(1 + u^2)^2} \cdot \dfrac{-2x}{(1 + x^2)^2}$. Now $1 + u^2 = 1 + \dfrac{1}{(1 + x^2)^2} = \dfrac{x^4 + 2x^2 + 2}{(1 + x^2)^2}$.

    So $\dfrac{dy}{du} = \dfrac{-2u}{(1 + u^2)^2} = \dfrac{-2}{1 + x^2} \cdot \dfrac{(1 + x^2)^4}{(x^4 + 2x^2 + 2)^2} = \dfrac{-2(1 + x^2)^3}{(x^4 + 2x^2 + 2)^2}$.

    Therefore $\dfrac{dy}{dx} = \dfrac{-2(1 + x^2)^3}{(x^4 + 2x^2 + 2)^2} \cdot \dfrac{-2x}{(1 + x^2)^2} = \dfrac{4x(1 + x^2)}{(x^4 + 2x^2 + 2)^2}$.

16. $\dfrac{dy}{dx} = \dfrac{15x^4 - 8x}{2(3x^5 - 4x^2)^{1/2}}.$

19. $2x^2 y \dfrac{dy}{dx} + 2xy^2 = 1 + \dfrac{dy}{dx},$ so $\dfrac{dy}{dx} = \dfrac{1 - 2xy^2}{2x^2 y - 1}.$

22. $\dfrac{dy}{dx} = \dfrac{(x^2 + \cos x)(1 + \cos x) - (x + \sin x)(2x - \sin x)}{(x^2 + \cos x)^2} = \dfrac{1 - x^2 - x \sin x + \cos x + x^2 \cos x}{(x^2 + \cos x)^2}$

28. $\dfrac{dy}{dx} = \dfrac{3(1 + \sqrt{x})^2}{2\sqrt{x}}(1 - 2\sqrt[3]{x})^4 + 4(1 - 2\sqrt[3]{x})^3(-\tfrac{2}{3}x^{-2/3})(1 + \sqrt{x})^3$

31. $\dfrac{dy}{dx} = (\sin^3 2x)(2)(\cos 3x)(-\sin 3x)(3) + (\cos^2 3x)(3 \sin^2 2x)(2 \cos 2x)$

  $= 6(\cos 3x \sin^2 2x)(\cos 3x \cos 2x - \sin 2x \sin 3x)$

  $= 6 \cos 3x \cos 5x \sin^2 2x$

34. $(x + y)^3 = (x - y)^2,$ so $3(x + y)^2 \left(1 + \dfrac{dy}{dx}\right) = 2(x - y) \left(1 - \dfrac{dy}{dx}\right);$

  thus $\dfrac{dy}{dx} = \dfrac{2(x - y) - 3(x + y)^2}{3(x + y)^2 + 2(x - y)} = \dfrac{2(x + y)^{3/2} - 3(x + y)^2}{3(x + y)^2 + 2(x + y)^{3/2}} = \dfrac{2 - 3\sqrt{x + y}}{2 + 3\sqrt{x + y}}.$

37. $1 = (2 \cos 2y)\dfrac{dy}{dx},$ so $\dfrac{dy}{dx} = \dfrac{1}{2 \cos 2y}.$ Because $\dfrac{dy}{dx}$ is undefined at $(1, \pi/4)$, there may well be a vertical tangent at that point; sure enough, $\dfrac{dx}{dy} = 0$ at $(1, \pi/4)$. So an equation of the tangent line is $x = 1$.

40. $V(x) = \tfrac{1}{3}\pi(36x^2 - x^3)$: $V'(x) = \pi x(24 - x)$. Now $\dfrac{dV}{dt} = \dfrac{dV}{dx} \cdot \dfrac{dx}{dt}$; when $x = 6$, $36\pi = -108\pi \dfrac{dx}{dt}$, so $dx/dt = -3$ (in./s) when $x = 6$.

43. $x \cot 3x = \dfrac{1}{3} \cdot \dfrac{3x}{\sin 3x} \to \dfrac{1}{3} \cdot 1 \cdot 1 = \dfrac{1}{3}$ as $x \to 0$.

46. $-1 \leq \sin u \leq 1$ for all u. So

$$-x^2 \leq x^2 \sin \dfrac{1}{x^2} \leq x^2$$

for all $x \neq 0$. But $x^2 \to 0$ as $x \to 0$, so the limit of the expression caught in the squeeze is also zero.

49. $g(x) = x^{-1/2}$, $f(x) = x^2 + 25$.

52. $g(x) = x^{10}$, $f(x) = \dfrac{x + 1}{x - 1}$; $h'(x) = -\dfrac{20(x + 1)^9}{(x - 1)^{11}}.$

55. $\dfrac{dV}{dS} \cdot \dfrac{dS}{dr} = \dfrac{dV}{dr}.$ Now $V = \dfrac{4}{3}\pi r^3$ and $S = 4\pi r^2$, so $\dfrac{dV}{dS} \cdot 8\pi r = 4\pi r^2$, and therefore $\dfrac{dV}{dS} = \dfrac{r}{2} = \dfrac{1}{4}\sqrt{\dfrac{S}{\pi}}.$

58. Current production per well: 200 (bbl/day). Number of new wells: x ($x \geq 0$). Production per well: $200 - 5x$. Total production:

$$T = T(x) = (20 + x)(200 - 5x), \qquad 0 \leq x \leq 40.$$

Now $T(x) = 4000 + 100x - 5x^2$, so $T'(x) = 100 - 10x$. $T'(x) = 0$ when $x = 10$. $T(0) = 4000$, $T(40) = 0$, and $T(10) = 4500$. So $x = 10$ maximizes $T(x)$. Answer: Ten new wells should be drilled, thereby increasing total production from 4000 bbl/day to 4500 bbl/day.

61. Let one sphere have radius r; the other, s. We seek the extrema of $A = 4\pi(r^2 + x^2)$ given $\tfrac{4}{3}\pi(r^3 + s^3) = V$, a constant. We illustrate here the **method of auxiliary variables:**

$$\dfrac{dA}{dr} = 4\pi \left(2r + 2s\dfrac{ds}{dr}\right);$$

the condition $dA/dr = 0$ yields $ds/dr = -r/s$. But we also know that $\tfrac{4}{3}\pi(r^3 + x^3) = V$; differentiation of both sides of this identity with respect to r yields

$$\frac{4}{3}\pi \left(3r^2 + 3s^2\frac{ds}{dr}\right) = 0, \text{ and so}$$

$$3r^2 + 3s^2\left(-\frac{r}{s}\right) = 0;$$

$$r^2 - rs = 0.$$

Therefore $r = 0$ or $r = s$. Also, ds/dr is undefined when $s = 0$. So we test these three critical points. If $r = 0$ or if $s = 0$, there is only one sphere, with radius $(3V/4\pi)^{1/3}$ and surface area $(36\pi V^2)^{1/3}$. If $r = s$, then there are two spheres of equal size, both with radius $\frac{1}{2}(3V/\pi)^{1/3}$ and surface area $(72\pi V^2)^{1/3}$. Therefore, for maximum surface area, make two equal spheres. For minimum surface area, make only one sphere.

64. Let x denote the length of the two sides of the corral that are perpendicular to the wall. There are two cases to consider.

Case 1: Part of the wall is used. Let y be the length of the side of the corral parallel to the wall. Then $y = 400 - 2x$, and we are to maximize the area

$$A = xy = x(400 - 2x), \qquad 150 \le x \le 200.$$

Then $A'(x) = 400 - 4x$; $A'(x) = 0$ when $x = 100$, but that value of x is not in the domain of A. Note that $A(150) = 15000$ and that $A(200) = 0$.

Case 2: All of the wall is used. Let y be the length of fence added to one end of the wall, so that the side parallel to the wall has length $100 + y$. Then $100 + 2y + 2x = 400$, so $y = 150 - x$. We are to maximize the area

$$A = x(100 + y) = x(250 - x), \qquad 0 \le x \le 150.$$

In this case $A'(x) = 0$ when $x = 125$. And in this case $A(150) = 15000$, $A(0) = 0$, and $A(125) = 15625$.

Answer: The maximum area is $15,625$ ft^2; to attain it, use all the existing wall and build a square corral.

67. Let x be the width of the base of the box, so that the base has length $2x$; let y be the height of the box. Then the volume of the box is $V = 2x^2y$, and for its total surface area to be 54 ft^2, we require $2x^2 + 6xy = 54$. Therefore the volume of the box is given by

$$V = V(x) = 2x^2\left(\frac{27 - x^2}{3x}\right) = \frac{2}{3}(27x - x^3), \qquad 0 < x \le 3\sqrt{3}.$$

Now $V'(x) = 0$ when $x^2 = 9$, so that $x = 3$. Also $V(0) = 0$, so even though $x = 0$ is not in the domain of V, the continuity of V implies that $V(x)$ is near zero for x near zero. Finally, $V(3\sqrt{3}) = 0$, so $V(3) = 36$ (ft^3) is the maximum possible volume of the box.

70. The square of the length of PQ is a function of x, $G(x) = (x - x_0)^2 + (y - y_0)^2$, which we are to maximize given the constraint $C(x) = y - f(x) = 0$. Now

$$\frac{dG}{dx} = 2(x - x_0) + 2(y - y_0)\frac{dy}{dx} \text{ and } \frac{dC}{dx} = \frac{dy}{dx} - f'(x).$$

When both vanish, $f'(x) = \dfrac{dy}{dx} = -\dfrac{x - x_0}{y - y_0}$. The line containing P and Q has slope

$$\frac{y - y_0}{x - x_0} = -\frac{1}{f'(x)},$$

and therefore this line is normal to the graph at Q.

73. As the diagram to the right suggests, we are to minimize the sum of the lengths of the two diagonals. Fermat's principle of least time may be used here, so we know that the angles at which the roads meet the shore are equal, and thus so are the tangents of those angles: $\frac{x}{1} = \frac{6-x}{2}$. It follows that the pier should be built two miles from the point on the shore nearest the first town. A short computation is sufficient to show that this actually yields the global minimum.

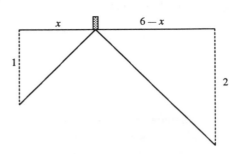

76. Here we have
$$R = R(\theta) = \frac{v^2\sqrt{2}}{16}(\cos\theta\,\sin\theta - \cos^2\theta)\ \text{ for }\ \pi/4 \le \theta \le \pi/2.$$
Now
$$R'(\theta) = \frac{v^2\sqrt{2}}{16}(\cos^2\theta - \sin^2\theta + 2\sin\theta\,\cos\theta);$$
$R'(\theta) = 0$ when $\cos 2\theta + \sin 2\theta = 0$, so that $\tan 2\theta = -1$. It follows that $\theta = 3\pi/8$ (67.5°). This yields the maximum range because $R(\pi/4) = 0 = R(\pi/2)$.

79. Use the iteration $x_{n+1} = x_n - \dfrac{(x_n)^5 - 75}{5(x_n)^4}$.
Results: $x_0 = 2.20379147$, $x_1 = 2.39896311$, $x_2 = 2.37206492$, $x_3 = 2.37144094$, $x_4 = 2.37144061$. Answer: 2.3714.

82. Linear interpolation gives $s_0 = -1/3$. We iterate using the formula
$$x_{n+1} = x_n - \frac{(x_n)^3 - 4x_n - 1}{3(x_n)^2 - 4}$$
to obtain the sequence -0.2525252525, -0.2541011930, -0.2541016884 of improving approximations. Answer: -0.2541.

85. Linear interpolation gives $x_0 = -0.5854549279$. If you got -0.9996955062 instead, it's because your calculator was set in degree mode. Change to radians and begin anew. Note also that a quick glance at the graphs of $y = -x$ and $y = \cos x$ shows that the equation $x + \cos x = 0$ has exactly one solution in the interval $[-2, 0]$. We use the Newton's method formula: We iterate
$$x \longleftarrow x - \frac{x + \cos x}{1 - \sin x}$$
and obtain -0.7451929664, -0.7390933178, -0.7390851332, -0.7390851332. Answer: -0.7391

88. Let $f(x) = 5(x + 1) - \cos x$. Then $f'(x) = 5 + \sin x$, and we iterate using the Newton's method formula
$$x \longleftarrow x - \frac{f(x)}{f'(x)}.$$
Linear interpolation yields the initial estimate -0.8809986055, and the succeeding approximations are -0.8712142932, -0.8712215145, and -0.8712215145. Answer: -0.8712.

91. Let $f(x) = x^5 - 3x^3 + x^2 - 23x + 19$. Then $f(-3) = -65$, $f(0) = 19$, $f(1) = -5$, and $f(3) = 121$. So there are at least three, and at most five, real solutions. Newton's method produces three real solutions, specifically $r_1 = -2.722493355$, $r_2 = 0.8012614801$, and $r_3 = 2.309976541$. If one divides the polynomial $f(x)$ by $(x - r_1)(x - r_2)(x - r_3)$, one obtains the quotient polynomial $x^2 + (0.38874466)x + 3.770552031$, which has no real roots—the quadratic formula yields the two complex roots $-0.194372333 \pm (1.932038153)i$. Consequently we have found all three real solutions.

94. We factor: $z^{3/2} - x^{3/2} = (z^{1/2})^3 - (x^{1/2})^3$
$$= (z^{1/2} - x^{1/2})(z + z^{1/2}x^{1/2} + x)$$

and $z - x = (z^{1/2})^2 - (x^{1/2})^2 = (z^{1/2} - x^{1/2})(z^{1/2} + x^{1/2})$. Therefore

$$\frac{z^{3/2} - x^{3/2}}{z - x} = \frac{z + z^{1/2}x^{1/2} + x}{z^{1/2} + x^{1/2}} \to \frac{3x}{2x^{1/2}} = \frac{3}{2}x^{1/2} \text{ as } z \to x.$$

97. The balloon has volume $V = \frac{4}{3}\pi r^3$ and surface area $A = 4\pi r^2$ where r is its radius and V, A, and r are all functions of time t. We are given $dV/dt = +10$, and we are to find dA/dt when $r = 5$.

$$\frac{dV}{dt} = 4\pi r^2 \frac{dr}{dt}, \text{ so } 10 = 4\pi \cdot 25 \cdot \frac{dr}{dt}.$$

$$\text{Thus } \frac{dr}{dt} = \frac{10}{100\pi} = \frac{1}{10\pi}.$$

$$\text{Also } \frac{dA}{dt} = 8\pi r \frac{dr}{dt}, \text{ and therefore}$$

$$\left.\frac{dA}{dt}\right|_{r=5} = 8\pi \cdot 4 \cdot \frac{1}{10\pi} = 4.$$

Answer: At 4 in.^{2}/s.

100. Let x denote the distance from plane A to the airport, y the distance from plane B to the airport, and z the distance between the two aircraft. Then

$$z^2 = x^2 + y^2 + (3 - 2)^2 = x^2 + y^2 + 1$$

and $dx/dt = -500$. Now

$$2z\frac{dz}{dt} = 2x\frac{dx}{dt} + 2y\frac{dy}{dt},$$

and when $x = 2$, $y = 2$. Therefore $z = 3$ at that time. Therefore,

$$3 \cdot (-600) = 2 \cdot (-500) + 2 \cdot \left.\frac{dy}{dt}\right|_{x=2},$$

and thus $\left.\dfrac{dy}{dt}\right|_{x=2} = -400$. Answer: Its speed is 400 mi/h.

103. The straight line through $P(x_0, y_0)$ and $Q(a, a^2)$ has slope $\dfrac{a^2 - y_0}{a - x_0} = 2a$, a consequence of the two-point formula for slope and the fact that the line is tangent to the parabola at Q. Hence $a^2 - 2ax_0 + y_0 = 0$. Think of this as a quadratic equation in the unknown a. It has two real solutions when the discriminant is positive: $(x_0)^2 - y_0 > 0$, and this establishes the conclusion in part (b). There are no real solutions when $(x_0)^2 - y_0 < 0$, and this establishes the conclusion in part (c). What if $(x_0)^2 - y_0 = 0$?

This seems a good spot to mention Descartes' Rule of Signs. In a sequence of nonzero numbers, such as

$$-3, \quad -5, \quad * \quad 2, \quad 4, \quad * \quad -1, \quad * \quad 6, \quad 6, \quad 6, \quad * \quad -2, \quad -2, \quad * \quad 12,$$

there are five *sign changes*, marked with asterisks. If some terms of such a sequence are zero, they are simply disregarded in counting the number of sign changes; thus the sequence

$$2, \quad 0, \quad 0, \quad * \quad -3, \quad -4, \quad 0, \quad 0, \quad 0, \quad * \quad 2, \quad 3, \quad * \quad -4, \quad 0,$$

has three sign changes.

Suppose that $p(x)$ is a polynomial with real coefficients, so that the polynomial equation $p(x) = 0$ has the form

$$a_0 x^n + a_1 x^{n-1} + \cdots + a_{n-1} x + a_n = 0.$$

Descartes' Rule of Signs is a classical theorem that states that the number of positive real roots of this equation is never greater than the number of sign changes in the sequence

$$a_0, \ a_1, \ a_2, \ \cdots, \ a_{n-1}, \ a_n,$$

and if less, then is less by an even number. For example, the equation

$$f(x) = x^5 + 3x^3 + x^2 - 23x + 19 = 0$$

of Problem 91 has the sequence 1, 0, −3, −23, 19 of coefficients, which has four sign changes. We already know that there are at least two and at most four positive solutions of the equation, and therefore there are either two or four. Now we consider the related equation

$$f(-x) = (-x)^5 - 3(-x)^3 + (-x)^2 - 23(-x) + 19 = 0,$$

which has the same number of positive roots as the original equation has negative roots. The sequence of coefficients in this equation is −1, 0, 3, 1, 23, 19; there is one sign change, so the original equation has exactly one negative root (because we already know of the existence of at least one).

4. $dy = -\dfrac{1}{(x - x^{1/2})^2}(1 - \frac{1}{2}x^{-1/2})\, dx$.

10. $dy = (x^2 \cos x + 2x \sin x)\, dx$.

16. $dy = -3(1 + \cos 2x)^{1/2}\, (\sin 2x)\, dx$

19. $f(x) \approx f(0) + f'(0)(x - 0) = 1 + 2x$

22. $f'(x) = -2(1 + 3x)^{-5/3}$, so $f(x) \approx f(0) + f'(0)(x - 0) = 1 - 2x$

25. Let $f(x) = x^{1/3}$. Then $f'(x) = \frac{1}{3}x^{-2/3}$. Take $a = 27$ and $x = 25$. Then $f(x) \approx f(a) + f'(a)(x - a)$, so

$$25^{1/3} = f(25)$$
$$\approx f(27) + f'(27)(25 - 27)$$
$$= 3 + \tfrac{1}{27}(-2) \approx 2.926.$$

28. Let $f(x) = x^{1/2}$; then $f'(x) = \frac{1}{2}x^{-1/2}$. Let $a = 81$ and let $x = 80$. Then

$$\sqrt{80} = f(x) \approx f(a) + f'(a)(x - a)$$
$$= \sqrt{81} + \frac{1}{2\sqrt{81}}(-1)$$
$$= \tfrac{161}{18} \approx 8.944.$$

34. $\dfrac{1}{2} - \dfrac{\pi}{180}\sqrt{3} \approx 0.46977.$

40. If C is the circumference of the circle and r its radius, then $C = 2\pi r$. Thus $dC = 2\pi\, dr$, and so $\Delta C \approx 2\pi \Delta r$. With $r = 10$ and $\Delta r = 0.5$, we obtain $\Delta C \approx 2\pi(0.5) = \pi \approx 3.1416$. This happens to be the exact value as well (because C is a linear function of r).

46. Because v is constant, R is a function of the angle of inclination θ alone, and hence

$$dR = \frac{1}{8}v^2(\cos 2\theta)\, d\theta.$$

With $\theta = \pi/4$, $d\theta = \pi/180$ (1°), and $v = 80$, we obtain

$$\Delta R \approx \frac{1}{8}(6400)(0)\frac{\pi}{180} = 0.$$

The true value of ΔR is approximately -0.2437.

49. Let V be the volume of the ball. Then $V_{\text{calc}} = \frac{4}{3}(1000\pi) \approx 4188.7902$ cubic inches. $\Delta V \approx 4\pi(10)^2\frac{1}{16} = 25\pi \approx 78.5398$ cubic inches (the true value of ΔV is approximately 79.0317).

52. With the notation of the preceding solution, we now require that

$$\frac{|dS|}{S} \leq 0.0001;$$

thus, at least approximately,

$$\frac{|4\pi r\, dr|}{|2\pi r^2|} \leq 0.0001.$$

Hence

$$2\left|\frac{dr}{r}\right| \leq 0.0001,$$

which implies that $|dr|/r \leq 0.00005$. Answer: With percentage error not exceeding 0.005%.

1. $f'(x) = -2x$; f is increasing on $(-\infty, 0)$ and decreasing on $(0, +\infty)$. Matching graph: (c).

4. $f'(x) = \frac{3}{4}x^2 - 3$; $f'(x) = 0$ when $x = \pm 2$; f is increasing on $(-\infty, -2)$ and on $(2, \infty)$, decreasing on $(-2, +2)$. Matching graph: (a).

7. $f(x) = 2x^2 + C$; $5 = f(0) = C$: $f(x) = 2x^2 + 5$.

10. $f(x) = 4\sqrt{x} + C$; $3 = f(0) = C$: $f(x) = 4\sqrt{x} + 3$.

13. $f'(x) = -4x$, so f is increasing on $(-\infty, 0)$ and decreasing on $(0, +\infty)$.

16. $f'(x) = 3x^2 - 12 = 3(x^2 - 4) = 3(x + 2)(x - 2)$. Hence f is increasing for $x > 2$ and for $x < -2$, decreasing for x in the interval $(-2, 2)$.

19. $f'(x) = 12x^3 + 12x^2 - 24x = 12x(x + 2)(x - 1)$, so the only points where $f'(x)$ can change sign are -2, 0, and 1.

22. $f'(x) = 6(x + 2)(x - 1)$, so f is increasing for $x < -2$ and for $x > 1$, decreasing for $-2 < x < 1$.

25. $f(0) = 0$, $f(2) = 0$, f is continuous for $0 \le x \le 2$, and $f'(x) = 2x - 2$ exists for $0 < x < 2$. To find the numbers c satisfying the conclusion of Rolle's theorem, we solve $f'(c) = 0$ to find that $c = 1$ is the only such number.

28. Here,
$$f'(x) = \frac{10}{3}x^{-1/3} - \frac{5}{3}x^{2/3} = \frac{10 - 5x}{3x^{1/3}};$$
$f'(x)$ exists for all x in $(0, 5)$ and f is continuous on the interval $0 \le x \le 5$ (the only point that might cause trouble is $x = 0$ but the limit of f and its value there are the same). Because $f(0) = 0 = f(5)$, there is a solution c of $f(x) = 0$ in $(0, 5)$, and clearly $c = 2$.

31. Because $f(1) = 2 \ne 0$, this function does not satisfy the hypotheses of Rolle's theorem. Neither does the conclusion hold, for $f'(x) = 4x^3 + 2x = 2x(2x^2 + 1)$ is never zero on $(0, 1)$.

34. First,
$$f'(x) = \frac{1}{2(x - 1)^{1/2}}$$
exists for all $x > 1$, so f satisfies the hypotheses of the mean value theorem for $2 \le x \le 5$. To find c, we solve
$$\frac{1}{2(c - 1)^{1/2}} = \frac{(4)^{1/2} - (1)^{1/2}}{5 - 2};$$
thus $2(c - 1)^{1/2} = 3$, and so $c = 13/4$. Note: $2 < c < 5$.

37. First, $f(x) = |x - 2|$ is not differentiable at $x = 2$, so does not satisfy the hypotheses of the mean value theorem on the given interval $1 \le x \le 4$. Wherever $f'(x)$ is defined, its value is 1 or -1, but
$$\frac{f(4) - f(1)}{4 - 1} = \frac{2 - 1}{3} = \frac{1}{3}$$
is never a value of $f'(x)$. So f satisfies neither the hypotheses nor the conclusion of the mean value theorem on the interval $1 \le x \le 4$.

40. The function $f(x) = 3x^{2/3}$ is continuous everywhere, but its derivative $f'(x) = 2x^{-1/3}$ does not exist at $x = 0$. Because $f'(x)$ does not exist for *all* x in $(-1, 1)$, this essential hypothesis of the mean value theorem is not satisfied. Moreover, $f'(x)$ is never zero (the average slope of the graph of f on the interval $-1 \le x \le 1$), so the conclusion of the theorem also fails to hold.

43. Let $g(x) = x^4 - 3x - 20$. Then $g(2) = -10 < 0$ and $g(3) = 52 > 0$. By the intermediate value property of continuous functions, $g(x)$ has at least one zero in $(2, 3)$. Its derivative $g'(x) = 4x^3 - 3$ is also continuous, and is zero only when $x = (3/4)^{1/3}$, approximately 0.91. So $g'(x)$ does not change

sign on the interval $2 \le x \le 3$. For $x \ge 2$, $g'(x) = 4x^3 - 3 \ge 4 \cdot 2^3 - 3 = 29 > 0$. Consequently g is increasing on that interval, and so $g(x)$ has at most one zero there. The conclusion is that $g(x) = 0$ has exactly one solution for $2 \le x \le 3$.

46. Proof: Suppose that $f'(x)$ is the constant K on the interval $a \le x \le b$. Let $g(x) = Kx + f(a) - Ka$. Then the graph of g is a straight line, and $g'(x) = K$ for all x. Consequently f and g differ by a constant on the interval $a \le x \le b$. But $g(a) = Ka + f(a) - Ka = f(a)$, so $g(x) = f(x)$ for all x in the interval. Therefore the graph of f is a straight line.

49. First note that $f'(x) = \frac{1}{2}x^{-1/2}$, and that the hypotheses of the mean value theorem are all satisfied for the given function on the given interval. Thus there does exist a number c between 100 and 101 such that

$$\frac{1}{2c^{1/2}} = \frac{f(101) - f(100)}{101 - 100} = \sqrt{101} - \sqrt{100}.$$

Therefore $1/(2\sqrt{c}) = \sqrt{101} - 10$, and thus we have shown that $\sqrt{101} = 10 + \dfrac{1}{2\sqrt{c}}$ for some number c in $(100, 101)$. Proof for part (b): If $0 \le \sqrt{c} \le 10$, then $0 \le c \le 100$; because $c > 100$, we see that $0 \le \sqrt{c} \le 10$ is impossible. If $10.5 \le \sqrt{c}$ then $110.25 \le c$, which is also impossible because $c < 110$. Therefore $10 < \sqrt{c} < 10.5$. Finally,

$$10 < \sqrt{c} < 10.5 \text{ implies that } 20 < 2\sqrt{c} < 21.$$

Consequently

$$\frac{1}{21} < \frac{1}{2c^{1/2}} < \frac{1}{20},$$

so

$$10 + \frac{1}{21} < \sqrt{101} < 10 + \frac{1}{20}.$$

The decimal expansion of $1/21$ begins $0.047619047619\ldots$, and so $10.0476 < \sqrt{101} < 10.05$.

52. The mean value theorem does not apply here because $f'(0)$ does not exist.

55. Use the definition of the derivative:

$$g'(0) = \lim_{h \to 0} \frac{g(0 + h) - g(0)}{h}$$

$$= \lim_{h \to 0} \frac{1}{2} + \frac{1}{h}h^2 \sin\left(\frac{1}{h}\right)$$

$$= \lim_{h \to 0} \frac{1}{2} + h \sin\left(\frac{1}{h}\right)$$

$$= \frac{1}{2} + 0 \quad \text{(by the squeeze law)}$$

$$= \frac{1}{2} > 0.$$

If $x \ne 0$ then

$$g'(x) = \frac{1}{2} + 2x \sin\left(\frac{1}{x}\right) - \cos\left(\frac{1}{x}\right).$$

Because $\cos(1/x)$ oscillates between $+1$ and -1 near $x = 0$ and $2x \sin(1/x)$ is near zero for x close to zero, it follows that every interval about $x = 0$ contains subintervals on which $g'(x) > 0$ and subintervals on which $g'(x) < 0$.

58. (a) Let $j(x) = \sin x - x + \frac{1}{6}x^3$. Then

$$j'(x) = \cos x - 1 + \frac{1}{2}x^2.$$

By Problem 57, $j'(x) > 0$ for all $x > 0$. Also, if $x > 0$, then $\dfrac{j(x) - 0}{x - 0} = j'(c)$ for some $c > 0$. Hence $j(x) > 0$ for all $x > 0$; that is, $\sin x > x - \frac{1}{6}x^3$ for all $x > 0$.

(b) By part (a) and Example 8,

$$x - \frac{1}{6}x^3 < \sin x < x$$

for all $x > 0$. So

$$\pi/36 - \frac{1}{6}(\pi/36)^3 < \sin \pi/36 < \pi/36;$$
$$0.0871557 < \sin 5° < 0.0872665;$$
$$\sin 5° \approx 0.087.$$

Section 4.4

1. $f'(x) = 2x - 4$; $x = 2$ is the only critical point. Because $f'(x) > 0$ for $x > 2$ and $f'(x) < 0$ for $x < 2$, it follows that $f(2) = 1$ is the global minimum value of $f(x)$.

4. $f'(x) = 3x^2 - 3 = 3(x + 1)(x - 1)$, so $x = 1$ and $x = -1$ are the only critical points. If $x < -1$ or if $x > 1$, then $f'(x) > 0$, while $f'(x) < 0$ on $(-1, 1)$. So $f(-1) = 7$ is a local maximum value while $f(1) = 3$ is a local minimum.

7. $f'(x) = -6(x - 5)(x + 2)$; $f'(x) < 0$ if $x < -2$ and if $x > 5$, while $f'(x) > 0$ for $-2 < x < 5$. Hence $f(-2) = -58$ is a local minimum value of $f(x)$ and $f(5) = 285$ is a local maximum value.

10. $f'(x) = 15x^2(x + 1)(x - 1)$, so $f'(x) > 0$ if $x < -1$ and if $x > 1$, while $f'(x) < 0$ on $(-1, 0)$ and on $(0, 1)$. Therefore $f(0) = 0$ is not an extremum of $f(x)$, but $f(-1) = 2$ is a local maximum value and $f(1) = -2$ is a local minimum value.

13. Here,

$$f'(x) = 2x - \frac{2}{x^2} = \frac{2(x^3 - 1)}{x^2} = \frac{2(x - 1)(x^2 + x + 1)}{x^2}.$$

Because $x^2 + x + 1 > 0$ for all x, the only critical point is $x = 1$; note that f is not defined at $x = 0$. Also $f'(x)$ has the sign of $x - 1$, so $f'(x) > 0$ for $x > 1$ and $f'(x) < 0$ for $0 < x < 1$ and for $x < 0$. Consequently $f(1) = 3$ is a local minimum value of $f(x)$. It is not a global minimum; check the behavior of $f(x)$ for x negative and near zero.

16. Because $f'(x) = \frac{1}{3}x^{-2/3}$, $f'(x) > 0$ for all x except for x except for $x = 0$, where f is continuous but $f'(x)$ is not defined. Consequently f has no extrema. Examination of the behavior of $f(x)$ and $f'(x)$ for x near zero makes it clear that the graph of f has a vertical tangent at $(0, 4)$.

19. $f'(x) = 3\sin^2 x \cos x$; $f'(x) = 0$ when $x = -\pi/2$, 0, or $\pi/2$. f is decreasing on $(-3, -\pi/2)$ and on $(\pi/2, 3)$, increasing on $(-\pi/2, \pi/2)$. So f has a [global] minimum at $(-\pi/2, -1)$ and a [global] maximum at $(\pi/2, 1)$.

22. $f'(x) = 3\tan^2 x \sec^2 x > 0$ on $(-1, 1)$; no extrema.

25. $f'(x) = 2\sec^2 x - 2\tan x \sec^2 x = (2\sec^2 x)(1 - \tan x)$. $f'(x) = 0$ when $x = \pi/4$; f is increasing on $(0, \pi/4)$ and decreasing on $(\pi/4, 1)$; maximum at $(\pi/4, 1)$.

28. We assume that the length turned upward is the same on each side—call it y. If the width of the gutter is x, then we have the constraint $xy = 18$, and we are to minimize the width $x + 2y$ of the strip. Its width is given by the function

$$f(x) = x + \frac{36}{x}, \qquad x > 0,$$

for which

$$f'(x) = 1 - \frac{36}{x^2}.$$

The only critical point in the domain of f is $x = 6$, and if $0 < x < 6$ then $f'(x) < 0$, while $f'(x) > 0$ for $x > 6$. Thus $x = 6$ yields the minimum value $f(6) = 12$ of the function f. Answer: The minimum possible width of the strip is 12 inches.

31. Base of box: x wide, $2x$ long. Height: y. Then the box has volume $2x^2y = 972$, so $y = 486x^{-2}$. Its total surface area is $A = 2x^2 + 6xy$, so we minimize

    $4x^2 + 5xy$

$$A = A(x) = 2x^2 + \frac{2916}{x}, \qquad x > 0.$$

Now

$$A'(x) = 4x - \frac{2916}{x^2},$$

so the only critical point of $A(x)$ occurs when $4x^3 = 2916$; that is, when $x = 9$. It is easy to verify that $A'(x) < 0$ for $0 < x < 9$ and that $A'(x) > 0$ for $x > 9$. Therefore $A(9)$ is the global minimum value of $A(x)$. Answer: The dimensions of the box are 9 inches wide, 18 inches long, 6 inches high.

34. If $(x, y) = (x, 4 - x^2)$ is a point on the parabola $y = 4 - x^2$, then the square of its distance from the point $(3, 4)$ is

$$h(x) = (x - 3)^2 + (4 - x^2 - 4)^2 = (x - 3)^2 + x^4.$$

We minimize the distance by minimizing its square:

$$h'(x) = 2(x - 3) + 4x^3;$$

$h'(x) = 0$ when $2x^3 + x - 3 = 0$. It is clear that $h'(1) = 0$, so $x - 1$ is a factor of $h'(x)$; $h'(x) = 0$ is equivalent to $(x - 1)(2x^2 + 2x + 3) = 0$. The quadratic factor in the last equation is always positive, so $x = 1$ is the only critical point of $h(x)$. Also $h'(x) < 0$ if $x < 1$, while $h'(x) > 0$ for $x > 1$, so $x = 1$ yields the global minimum value $h(1) = 5$ for $h(x)$. When $x = 1$ we have $y = 3$, so the point on the parabola $y = 4 - x^2$ closest to $(3, 4)$ is $(1, 3)$, at distance $\sqrt{5}$ from it.

37. Let the square base of the box have edge length x and let its height be y, so that its total volume is $x^2y = 62.5$ and the surface area of this box-without-top will be $A = x^2 + 4xy$. So

$$A = A(x) = x^2 + \frac{250}{x}, \qquad x > 0.$$

Now

$$A'(x) = 2x - \frac{250}{x^2},$$

so $A'(x) = 0$ when $x^3 = 125$: $x = 5$. In this case, $y = 2.5$. Also $A'(x) < 0$ if $0 < x < 5$ and $A'(x) > 0$ if $x > 5$, so we have found the global minimum for $A(x)$. Answer: Square base of edge length 5 inches, height 2.5 inches.

40. If the print width is x and its height is y (in inches), then the page area is $A = (x + 2)(y + 4)$. We are to minimize A given $xy = 30$. Because $y = 30/x$,

$$A = A(x) = 4x + 38 + \frac{60}{x}, \qquad x > 0.$$

Now

$$A'(x) = 4 - \frac{60}{x^2};$$

$A'(x) = 0$ when $x = \sqrt{15}$. But $A'(x) > 0$ for $x > \sqrt{15}$ while $A'(x) < 0$ for $x < \sqrt{15}$. Therefore $x = \sqrt{15}$ yields the global minimum value of $A(x)$, which is $38 + 8\sqrt{15}$, approximately 68.98 square inches.

43. If the dimensions of the rectangle are x by y, and the line segment bisects the side of length x, then the square of the length of the segment is

$$f(x) = \left(\frac{x}{2}\right)^2 + y^2 = \frac{x^2}{4} + \frac{4096}{x^2}, \qquad x > 0,$$

because $y = 64/x$. Now

$$f'(x) = \frac{x}{2} - \frac{8192}{x^3}.$$

When $f'(x) = 0$, we must have $x = +8\sqrt{2}$, so that $y = 4\sqrt{2}$. We have found the minimum of f because if $0 < x < 8\sqrt{2}$ then $f'(x) < 0$, while $f'(x) > 0$ if $x > 8\sqrt{2}$. The minimum length satisfies $L^2 = f(8\sqrt{2})$, so that $L = 8$ centimeters.

46.   By similar triangles, $y/1 = 8/x$, and

$$L_1 + L_2 = L = \left((x+1)^2 + (y+8)^2\right)^{1/2}.$$

We minimize L by minimizing

$$f(x) = L^2 = (x+1)^2 + \left(8 + \frac{8}{x}\right)^2, \qquad x > 0.$$

$$f'(x) = 2 + 2x - \frac{128}{x^3} - \frac{128}{x^2};$$

$f'(x) = 0$ when $2x^3 + 2x^4 - 128 - 128x = 0$, which leads to the equation

$$(x+1)(x-4)(x^2 + 4x + 16) = 0.$$

The only relevant solution is $x = 4$. Because $f'(x) < 0$ for x in the interval $(-1, 4)$ and $f'(x) > 0$ if $x > 4$, we have indeed found the global minimum of f. The corresponding value of y is 2, and the length of the shortest ladder is $L = 5\sqrt{5}$ feet, approximately 11 ft 2 in.

49.   Let z be the length of the segment from the top of the tent to the midpoint of one side of its base. Then $x^2 + y^2 = z^2$. The total surface area of the tent is

$$A = 4x^2 + (4)(\tfrac{1}{2})(2x)(z) = 4x^2 + 4xz = 4x^2 + 4x(x^2 + y^2)^{1/2}.$$

Because the (fixed) volume V of the tent is given by

$$V = \frac{1}{3}(4x^2)(y) = \frac{4}{3}x^2 y,$$

we have $y = 3V/(4x^2)$, so

$$A = A(x) = 4x^2 + \frac{1}{x}(16x^6 + 9V^2)^{1/2}.$$

After simplifications, the condition $dV/dx = 0$ takes the form

$$8x(16x^6 + 9V^2)^{1/2} - \frac{1}{x^2}(16x^6 + 9V^2) + 48x^4 = 0,$$

which has solution $x = 2^{-7/6}\sqrt[3]{3V}$. Because this is the only positive solution of the equation, and because it is clear that neither large values of x nor values of x near zero will yield small values of the surface area, this is the desired value of x.

Section 4.5

1. $f(x) \to +\infty$ as $x \to +\infty$, $f(x) \to -\infty$ as $x \to -\infty$. Matching graph: 4.5.11(c).

4. $f(x) \to -\infty$ as $x \to +\infty$, $f(x) \to -\infty$ as $x \to -\infty$. Matching graph: 4.5.11(b).

10. $f'(x) = 3x^2 + 12x + 9 = 3(x + 1)(x + 3)$ is positive for $x > -1$ and for $x < -3$, negative for $-3 < x < -1$. So there is a local maximum at $(-3, 0)$ and a local minimum at $(-1, -4)$. There are intercepts at $(-3, 0)$ and $(0, 0)$.

16. Here, $f'(x) = \dfrac{10 - 5x}{3x^{1/3}}$ can change sign only at $x = 2$ and at $x = 0$. The function is increasing for $0 < x < 2$, decreasing for $x < 0$ and for $x > 2$. Thus there is a local maximum at $(2, f(2))$ and a local minimum at $(0, 0)$. Note that $f'(0)$ does not exist, but that f is continuous at $x = 0$.

22. The graph is a parabola, opening downward, vertical axis, vertex at $(-5/12, 169/24)$.

28. $f(x) = (x^2 - 1)^2$; $f'(x) = 4x(x + 1)(x - 1)$. So $f(x)$ is increasing for $-1 < x < 0$ and for $x > 1$, decreasing if $x < -1$ or if $0 < x < 1$. There are local minima at $(-1, 0)$ and at $(1, 0)$ and a local maximum at $(0, 1)$. These points are also all the intercepts.

34. Here we have
$$f'(x) = -\frac{1}{3x^{2/3}},$$
which is negative for all $x \neq 0$. Though $f'(0)$ is not defined, f is continuous at $x = 0$; careful examination of the behavior of f and f' near zero shows that the graph has a vertical tangent at $(0, 1)$; there are no extrema.

40. The graph is shown below.

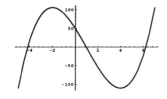

43. The graph is shown below.

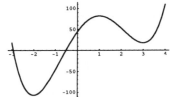

46. The graph of $y = x^3 - 3x - 2$:

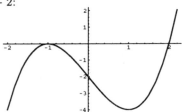

Section 4.6

1. $f'(x) = 8x^3 - 9x^2 + 6$, $f''(x) = 24x^2 - 18x$, $f'''(x) = 48x - 18$

4. $g'(t) = 2t + \frac{1}{2}(t + 1)^{-1/2}$, $g''(t) = 2 - \frac{1}{4}(t + 1)^{-3/2}$, $g'''(t) = \frac{3}{8}(t + 1)^{-5/2}$

10. $h'(z) = \dfrac{8z}{(z^2 + 4)^2}$, $h''(z) = \dfrac{32 - 24z^2}{(z^2 + 4)^3}$, $h'''(z) = \dfrac{96z^3 - 384z}{(z^2 + 4)^4}$

16. Given: $x^2 + y^2 = 4$.
$$2x + 2yy'(x) = 0, \quad \text{so} \quad y'(x) = -\frac{x}{y}.$$
$$y''(x) = -\frac{y - xy'(x)}{y^2} = -\frac{y + (x^2/y)}{y^2} = -\frac{y^2 + x^2}{y^3} = -\frac{4}{y^3}.$$

22. $\sin^2 x + \cos^2 y = 1$: $2\sin x \cos x = 2y'(x)\sin y \cos y = 0$; $y'(x) = \dfrac{\sin x \cos x}{\sin y \cos y}$.

$\dfrac{d^2 y}{dx^2}$ can be simplified (with the aid of the original equation) to

$$\frac{\cos^2 x \sin^2 y - \sin^2 x \cos^2 y}{\sin^3 y \cos^3 y} \equiv 0 \ \text{ if } y \text{ is not an integral multiple of } \pi/2.$$

28. Critical point: $(7.5, -1304.69)$; inflection points: $(0, -250)$ and $(5, -875)$

34. $f'(x) = 3x(x - 2)$; $f''(x) = 6(x - 1)$. There are critical points at $(0, 0)$ and at $(2, -4)$. Now $f''(0) = -6 < 0$, so there is a local maximum at $(0, 0)$; $f''(2) = 6 > 0$, so there is a local minimum at $(2, -4)$. The only possible inflection point is $(1, -2)$, and it is indeed an inflection point because f'' changes sign there.

40. $f'(x) = x^2(x + 2)(5x + 6)$ and $f''(x) = 4x(5x^2 + 12x + 6)$. So the critical points occur where $x = 0$, $x = -2$, and $x = -6/5$. Now $f''(0) = 0$, so the second derivative test fails here, but $f'(x) > 0$ for x near zero but $x \neq 0$, so $(0, 0)$ is not an extremum for the graph of f. Next, $f''(-2) = -16 < 0$, so $(-2, 0)$ is a local maximum point; $f''(-6/5) = 5.76 > 0$, so $(-6/5, -3456/3125)$ is a local minimum point. The possible inflection points occur at

$$x = 0, \quad x = \frac{1}{5}\left(-6 + \sqrt{6}\right), \quad \text{and} \quad \frac{1}{5}\left(-6 - \sqrt{6}\right).$$

In decimal form these are $x = 0$, $x \approx -0.710$, and $x \approx -1.690$. Because $f''(-2) = -16 < 0$, $f''(-1) = 4 > 0$, $f''(-0.5) = -2.5 < 0$, and $f''(1) = 92 > 0$, each of the three numbers displayed above is the abscissa of an inflection point of the graph of f.

46. $f(x) = \sin^3 x$, $-\pi < x < \pi$:

$$f'(x) = 3\sin^2 x \cos x,$$
$$f''(x) = 6\sin x \, \cos^2 x - 3\sin^3 x = 3(2\cos^2 x - \sin^2 x)\sin x.$$

Minimum at $(-\pi/2, -1)$; maximum at $(\pi/2, 1)$; inflection points at $x = 0$ and at the four solutions of $\tan^2 x = 2$ in $(-\pi, \pi)$.

52. We assume that the length turned upward is the same on each side—call it y. If the width of the gutter is x, then we have the constraint $xy = 18$, and we are to minimize the width $x + 2y$ of the strip. Its width is given by the function

$$f(x) = x + \frac{36}{x}, \qquad x > 0,$$

for which

$$f'(x) = 1 - \frac{36}{x^2} \quad \text{and} \quad f''(x) = \frac{72}{x^3}.$$

The only critical point in the domain of f is $x = 6$, and $f''(x) > 0$ on the entire domain of f. Consequently the graph of f is concave upward for all $x > 0$. Because f is continuous for such x, $(6, 12)$ is the global minimum of f.

55. Base of box: x wide, $2x$ long. Height: y. Then the box has volume $2x^2 y = 972$, so $y = 486x^{-2}$. Its total surface area is $A = 2x^2 + 6xy$, so we minimize

$$A = A(x) = 2x^2 + \frac{2916}{x}, \qquad x > 0.$$

Now

$$A'(x) = 4x - \frac{2916}{x^2} \quad \text{and} \quad A''(x) = 4 + \frac{5832}{x^3}.$$

The only critical point of A occurs when $x = 9$, and $A''(x)$ is always positive. So the graph of $y = A(x)$ is concave upward for all $x > 0$; consequently, $(9, A(9))$ is the global minimum point of A. Answer: The dimensions of the box are 9 inches wide, 18 inches long, and 6 inches high.

58. Let x denote the length of each side of the square base of the solid and let y denote its height. Then its total volume is $x^2 y = 1000$. We are to minimize its total surface area $A = 2x^2 + 4xy$. Now $y = 1000/(x^2)$, so

$$A = A(x) = 2x^2 + \frac{4000}{x}, \qquad x > 0.$$

Therefore
$$A'(x) = 4x - \frac{4000}{x^2} \quad \text{and} \quad A''(x) = 4 + \frac{8000}{x^3}.$$

The only critical point occurs when $x = 10$, and $A''(x) > 0$ for all x in the domain of A, so $x = 10$ yields the global minimum value of $A(x)$. In this case, $y = 10$ as well, so the solid is indeed a cube.

61. Let x denote the radius and y the height of the cylinder (in inches). Then its cost (in cents) is $C = 8\pi x^2 + 4\pi xy$, and we also have the constraint $\pi x^2 y = 100$. So
$$C = C(x) = 8\pi x^2 + \frac{400}{x}, \qquad x > 0.$$

Now
$$C'(x) = 16\pi x - \frac{400}{x^2} \quad \text{and} \quad C''(x) = 16\pi + \frac{800}{x^3}.$$

The only critical point in the domain of C is $x = \sqrt[3]{25/\pi}$ (about 1.9965 inches) and, consequently, when $y = \sqrt[3]{1600/\pi}$ (about 7.9859 inches). Because $C''(x) > 0$ for all x in the domain of C, we have indeed found the dimensions that minimize the cost of the can. For simplicity, note that $y = 4x$ at the minimum: The height of the can is twice its diameter.

64. Given: $f(x) = 3x^4 - 4x^3 - 5$. Then
$$f'(x) = 12x^3 - 12x^2 = 12x^2(x - 1) \quad \text{and} \quad f''(x) = 36x^2 - 24x = 12x(3x - 2).$$

So the graph of f is increasing for $x > 1$ and decreasing for $x < 1$ (even though there's a horizontal tangent at $x = 0$), concave upward for $x < 0$ and $x > 2/3$, concave downward on $(0, 2/3)$. There is a global minimum at $(1, -6)$, inflection points at $(0, -5)$ and at $(2/3, -151/27)$. The x-intercepts are approximately -0.906212 and 1.682971.

70. Given: $f(x) = (x - 1)^2(x + 2)^3$. Then
$$f'(x) = (x - 1)(x + 2)^2(5x + 1) \quad \text{and} \quad f''(x) = 2(x + 2)(10x^2 + 4x - 5).$$

The zeros of $f''(x)$ are $x = -2$, $x \approx 0.535$, and $x \approx -0.935$. It follows that $(1, 0)$ is a local minimum (from the second derivative test), that $(-0.2, 8.39808)$ is a local maximum, and that $(-2, 0)$ is not an extremum. Also, the second derivative changes sign at each of its zeros, so each of these three zeros is the abscissa of an inflection point on the graph.

76. Given: $f(x) = x^{1/3}(6 - x)^{2/3}$. Then
$$f'(x) = \frac{2 - x}{x^{2/3}(6 - x)^{1/3}} \quad \text{and} \quad f''(x) = -\frac{8}{x^{5/3}(6 - x)^{4/3}}.$$

If $|x|$ is large, then $(6 - x)^{2/3} \approx x^{2/3}$, so $f(x) \approx x$ for such x. This aids in sketching the graph, which has a local maximum where $x = 2$, a local minimum at $(6, 0)$, vertical tangents at $(6, 0)$ and at the origin. It is increasing for $x < 2$ and for $x > 6$, decreasing on the interval $(2, 6)$, concave upward for $x < 0$, and concave downward on $(0, 6)$ and for $x > 6$. All the intercepts have been mentioned, too.

82. Figure 4.6.33 (a)

85. $\dfrac{dz}{dx} = \dfrac{dz}{dy} \cdot \dfrac{dy}{dx}$. So $\dfrac{d^2z}{dx^2} = \dfrac{dz}{dy} \cdot \dfrac{d^2y}{dx^2} + \dfrac{dy}{dx} \cdot \dfrac{d^2z}{dy^2} \cdot \dfrac{dy}{dx}$.

88. If $f(x) = Ax^4 + Bx^3 + Cx^2 + Dx + E$, then both $f'(x)$ and $f''(x)$ are continuous for all x, and $f''(x) = 12Ax^2 + 6Bx + 2C$. In order for $f''(x)$ to change sign, we must have $f''(x) = 0$. If so, then (because $f''(x)$ is a quadratic polynomial) either the graph of $f''(x)$ crosses the x-axis in two places or is tangent to it at a single point. In the first case, $f''(x)$ changes sign twice, so there are two points of inflection on the graph of f. In the second case, $f''(x)$ does not change sign, so f has no inflection points. Therefore the graph of a polynomial of degree four has either exactly two inflection points or else none at all.

1. $\dfrac{x}{x+1} = \dfrac{1}{1+(1/x)} \to 1$ as $x \to +\infty$.

4. The numerator approaches -2 as $x \to -1$, while the denominator approaches zero. Therefore this limit does not exist.

7. The numerator is equal to the denominator for all $x \neq -1$, so the limit is 1.

10. Divide each term in numerator and denominator by $x^{3/2}$, the highest power of x that appears in any term. The numerator then becomes $2x^{-1/2} + x^{-3/2}$, which approaches 0 as $x \to +\infty$; the denominator becomes $x^{-1/2} - 1$, which approaches -1 as $x \to +\infty$. Therefore the limit is 0.

13. $\dfrac{4x^2 - x}{x^2 + 9} = \dfrac{4 - (1/x)}{1 + (9/x^2)} \to 4$ as $x \to +\infty$, so the limit is $\sqrt{4} = 2$.

16. $\displaystyle \lim_{x \to -\infty} \left(2x - \sqrt{4x^2 - 5x}\right) = \lim_{x \to -\infty} \dfrac{4x^2 - (4x^2 - 5x)}{2x + \sqrt{4x^2 - 5x}} = \lim_{x \to -\infty} \dfrac{5x}{2x + \sqrt{4x^2 - 5x}}$

$\displaystyle = \lim_{x \to -\infty} \dfrac{5}{2 + \left(\dfrac{\sqrt{4x^2 - 5x}}{-\sqrt{x^2}}\right)} = \lim_{x \to -\infty} \dfrac{5}{2 - \sqrt{4 - (5/x)}} = -\infty.$

22. Matches 4.7.9 (c)

28. Matches 4.7.9 (e)

31. Here we have
$$f'(x) = -\frac{6}{(x+2)^3} \quad \text{and} \quad f''(x) = \frac{18}{(x+2)^4}.$$
The graph is increasing and concave upward for $x < -2$, decreasing and concave upward for $x > -2$. There are no extrema and no inflection points; the y-intercept is $(0, \frac{3}{4})$. The line $x = -2$ is a vertical asymptote and the x-axis is a horizontal asymptote.

34. Here,
$$f'(x) = -\frac{2}{(x-1)^2} \quad \text{and} \quad f''(x) = \frac{4}{(x-1)^3}.$$
The graph is decreasing and concave downward if $x < 1$, decreasing and concave upward if $x > 1$. There are no extrema and no inflection points; the y-intercept is $(0, -1)$ and the only x-intercept is $(-1, 0)$. The line $x = 1$ is a vertical asymptote and the line $y = 1$ is a horizontal asymptote.

37.
$$f(x) = \frac{1}{x^2 - 9} \qquad f'(x) = -\frac{2x}{(x^2 - 9)^2} \quad \text{and} \quad f''(x) = \frac{6(x^2 + 3)}{(x^2 - 9)^3}.$$
The graph is increasing for $x < -3$ and for $-3 < x < 0$; the graph is decreasing for $0 < x < 3$ and for $3 < x$. It is concave upward if $x < -3$ and if $x > 3$, concave downward on $(-3, 3)$. The only extremum is the local maximum and y-intercept at the point $(0, -\frac{1}{9})$, and there are no points of inflection. The x-axis is a horizontal asymptote and the lines $x = -3$ and $x = 3$ are vertical asymptotes.

40.
$$\frac{2x^2 + 1}{x^2 - 2x} = f(x) \qquad f'(x) = -\frac{2(2x - 1)(x + 1)}{x^2(x - 2)^2} \quad \text{and} \quad f''(x) = \frac{2\left(4x^3 + 3x^2 - 6x + 4\right)}{(x^2 - 2x)^3}.$$
The zeros of $f'(x)$ occur at -1 and 0.5, and the only zero of $f''(x)$ is at $x \approx -1.8517$. The graph is decreasing for $x < -1$, for $0.5 < x < 2$, and for $x > 2$; increasing for $-1 < x < 0$ and for $0 < x < 0.5$. It is concave downward for $x < -1.85^*$ and on $(0, 2)$, concave upward for $x > 2$ and on $(-1.85^*, 0)$. The line $y = 2$ is a horizontal asymptote and the lines $x = 0$ and $x = 2$ are vertical asymptotes. There is a local minimum at $(-1, 1)$, a local maximum at $(0.5, -2)$, and a point of inflection at $(-1.85^*, 1.10^*)$. (*Coordinates approximate.)

43.

$$f'(x) = \frac{x(x-2)}{(x-1)^2} \quad \text{and} \quad f''(x) = \frac{2}{(x-1)^3}.$$

The graph is increasing for $x < 0$ and for $x > 2$, decreasing on $(0, 1)$ and on $(1, 2)$. It is concave upward if $x > 1$, concave downward if $x < 1$. The origin is a local maximum and the only intercept, the only other extremum is a local minimum at $(2, 4)$, and there are no inflection points. The lines $y = x + 1$ and $x = 1$ are asymptotes.

46.

$$f'(x) = -\frac{2x}{(x^2-4)^2} \quad \text{and} \quad f''(x) = \frac{6x^2+8}{(x^2-4)^3}.$$

The graph is increasing for $x < -2$ and on $(-2, 0)$, decreasing on $(0, 2)$ and for $x > 2$. It is concave upward if $|x| > 2$ and concave downward if $|x| < 2$. The only intercept is at the point $(0, -0.25)$, which is also a local maximum; there are no other extrema and no inflection points. The two vertical lines $x = -2$ and $x = 2$ are asymptotes, as is the x-axis.

49.

$$f(x) = \frac{1}{x^2-x-2} = \frac{1}{(x-1)(x-2)} : \quad f'(x) = -\frac{2x-1}{(x+1)^2(x-2)^2} \quad \text{and} \quad f''(x) = \frac{6(x^2-x+1)}{(x+1)^3(x-2)^3}.$$

It is useful to notice the discontinuities at $x = -1$ and at $x = 2$, that $f(x) \to 0$ as $x \to +\infty$ and as $x \to -\infty$, and that $f(x) > 0$ for $x > 2$ and for $x < -1$ while $f(x) < 0$ on the interval $(-1, 2)$. The denominator in $f'(x)$ is never negative, so the graph is decreasing for $\frac{1}{2} < x < 2$ and for $2 < x$ and increasing for $x < -1$ and for $-1 < x < \frac{1}{2}$. The graph is concave upward for $x > 2$ and for $x < -1$, concave downward on $(-1, 2)$. There are no inflection points or intercepts, but there is a local maximum at $(\frac{1}{2}, -\frac{4}{9})$. The x-axis is a horizontal asymptote and the lines $x = -1$ and $x = 2$ are vertical asymptotes.

52.

$$f'(x) = -\frac{x^2+1}{(x^2-1)^2} \quad \text{and} \quad f''(x) = \frac{2x(x^2+3)}{(x^2-1)^3}.$$

The graph is decreasing on $(-\infty, -1)$, on $(-1, 1)$, and on $(1, \infty)$; it is concave upward for $x > 1$ and for $-1 < x < 0$, concave downward for $0 < x < 1$ and for $x < -1$. The only intercept is $(0, 0)$, which is also an inflection point. Note that $f(x) \to 0$ as $|x| \to \infty$, so the x-axis is a horizontal asymptote; $f(x) \to +\infty$ as $x \to -1^+$ and as $x \to +1^+$, while $f(x) \to -\infty$ as $x \to -1^-$ and as $x \to +1^-$, so the lines $x = -1$ and $x = 1$ are vertical asymptotes. There are no extrema.

55. Sketch the parabola $y = x^2$, but modify it by changing its behavior near $x = 0$: Let $y \to +\infty$ as $x \to 0^+$ and let $y \to -\infty$ as $x \to 0^-$. Using calculus, we compute

$$f'(x) = \frac{2(x^3-1)}{x^2} \quad \text{and} \quad f''(x) = \frac{2(x^3+2)}{x^3}.$$

It follows that the graph is decreasing for $0 < x < 1$ and for $x < 0$, increasing for $x > 1$. It is concave upward for $x < -\sqrt[3]{2}$ and also for $x > 0$, concave downward for $-\sqrt[3]{2} < x < 0$. The only intercept is at $(-\sqrt[3]{2}, 0)$; this is also the only inflection point. There is a local minimum at $(1, 3)$. The y-axis is a vertical asymptote.

Chapter 4 Miscellaneous

1. $dy = \frac{3}{2}(4x - x^2)^{1/2}(4 - 2x)\, dx$

4. $dy = 2x \cos(x^2)\, dx$

7. Let $f(x) = x^{1/2}$; $f'(x) = \frac{1}{2}x^{-1/2}$. Then

$$\sqrt{6401} = f(6400 + 1) \approx f(6400) + 1 \cdot f'(6400)$$

$$= 80 + \frac{1}{160} = \frac{12801}{160} = 80.00625.$$

(Actually, $\sqrt{6401} \approx 80.00624976$.)

10. $\sqrt[3]{999} = \sqrt[3]{1000} - 1 \cdot \frac{1}{3}(1000)^{-2/3} = 10 - \frac{1}{300} = \frac{2999}{300} \approx 9.996667$.

16. With $f(x) = x^{1/10}$, $f'(x) = \frac{1}{10}x^{-9/10}$, $x = 1024$, and $\Delta x = -24$, we obtain

$$(1000)^{1/10} = f(x + \Delta x) \approx f(x) + f'(x)\,\Delta x$$

$$= (1024)^{1/10} + (-24)\left(\frac{1}{10}\right)(1024)^{-9/10}$$

$$= 2 - \frac{3}{640} = \frac{1277}{640} \approx 1.9953.$$

22. Here, $dL = (-13)(10^{30})E^{-14}\,dE$. We take $E = 110$ and $\Delta E = +1$ and obtain

$$dL \approx (-13)(10^{30})\left(110^{-14}\right)(+1) \approx -342 \quad \text{hours}.$$

The actual decrease is $L(110) - L(111) \approx 2896.6 - 2575.1 \approx 321.5$ (hours).

28. $\dfrac{f(4) - f(0)}{4 - 0} = f'(c)$: $\quad \dfrac{2 - 0}{4} = \frac{1}{2}c^{-1/2}$; $c^{-1/2} = 1$; $c = 1$.

31. $f'(x) = 15(x^4 - x^2 + 4) > 0$ for all x; $f''(x) = 30x(2x - 1)$, so the graph is concave upward for $x < 0$, concave downward on the interval $(0, 1/2)$, and concave upward for $x > 1/2$. Therefore there are no extrema, $(0, 0)$ is the only intercept, and there are inflection points at $(0, 0)$ and $(1/2, 943/32)$.

34. Let $g(x) = x^5 + x - 5$. Then $g(2) = 29 > 0$ while $g(1) = -3 < 0$. Because $g(x)$ is a polynomial, it has the intermediate value property. Therefore the equation $g(x) = 0$ has at least one solution in the interval $1 \le x \le 2$. Moreover, $g'(x) = 5x^4 + 1$, so $g'(x) > 0$ for all x. Consequently g is increasing on the set of all real numbers, and so takes on each value—including zero—at most once. We may conclude that the equation $g(x) = 0$ has exactly one solution, and hence that the equation $x^5 + x = 5$ has exactly one solution. (The solution is approximately 1.299152792.)

40. $g'(x) = -\dfrac{2x}{(x^2 + 9)^2}$, $g''(x) = \dfrac{6x^2 - 18}{(x^2 + 9)^3}$, and $g'''(x) = \dfrac{216x - 24x^3}{(x^2 + 9)^4}$.

46. $\dfrac{dy}{dx} = \dfrac{3y - 4x}{10y - 3x}$; $\dfrac{d^2y}{dx^2} = -\dfrac{1550}{(10y - 3x)^3}$.

52. $(x^2 - y^2)^2 = 4xy$:

$$2(x^2 - y^2)\left(2x - 2y\frac{dy}{dx}\right) = 4x\frac{dy}{dx} + 4y$$

$$(x^2 - y^2)\cdot x - (x^2 - y^2)\cdot y\frac{dy}{dx} = x\frac{dy}{dx} + y$$

$$(x + x^2y - y^3)\frac{dy}{dx} = x^3 - xy^2 - y$$

$$\frac{dy}{dx} = \frac{x(x^2 - y^2) - y}{x + y(x^2 - y^2)} = \frac{x^3 - xy^2 - y}{x + x^2y - y^3}.$$

$$\frac{d^2y}{dx^2} = \frac{(x + x^2y - y^3)\left(3x^2 - x2y\dfrac{dy}{dx} - y^2 - \dfrac{dy}{dx}\right) - (x^3 - xy^2 - y)\left(1 + x^2\dfrac{dy}{dx} + 2xy - 3y^2\dfrac{dy}{dx}\right)}{(x + x^2y - y^3)^2},$$

which upon simplification and substitution for dy/dx becomes:

$$\frac{d^2y}{dx^2} = \frac{3xy(2 - xy)}{(x + x^2y - y^3)^3}.$$

55. $f'(x) = 2x^3(3x^2 - 4)$ and $f''(x) = 6x^2(5x^2 - 4)$.

58. $f'(x) = \dfrac{3}{(x+2)^2}$ and $f''(x) = -\dfrac{6}{(x+2)^3}$ There are no critical points and no inflection points. The graph is increasing except at the discontinuity at $x = -2$. It is concave upward for $x < -2$ and concave downward for $x > -2$. The vertical line $x = -2$ and the horizontal line $y = 1$ are asymptotes.

61. $f(x) = \dfrac{2x^2}{(x-2)(x+1)}$, $f'(x) = -\dfrac{2x(x+4)}{(x-2)^2(x+1)^2}$, and $f''(x) = \dfrac{4(x^3 + 6x^2 + 4)}{(x^2 - x - 2)^3}$.

64. Here we have $f(x) = x^2(x^2 - 2)$, $f'(x) = 4x(x+1)(x-1)$, and $f''(x) = 4(3x^2 - 1)$. So there are intercepts at $(-\sqrt{2}, 0)$, $(0, 0)$, and $(\sqrt{2}, 0)$. The graph is increasing on the intervals $(1, \infty)$ and $(-1, 0)$, decreasing on the intervals $(-\infty, -1)$ and $(0, 1)$. It is concave upward where $x^2 > 1/3$ and concave downward where $x^2 < 1/3$. There are global minima at $(-1, -1)$ and $(1, -1)$ and a local maximum at the origin. There are inflection points at the two points where $x^2 = 1/3$.

67. $f'(x) = 3x(4 - x)$ and $f''(x) = 6(2 - x)$. The abscissas of the x-intercepts are approximately -1.18014, 1.48887, and 5.69127.

70. We have $f(x) = x^2(x^2 - 12)$, $f'(x) = 4x(x^2 - 6)$, and $f''(x) = 12(x^2 - 2)$. The analysis, results, and graph are qualitatively the same as in the solution of Problem 64.

73. The given function $f(x)$ is expressed as a fraction with constant numerator, so we maximize $f(x)$ by minimizing its denominator $(x+1)^2 + 1$. It is clear that $x = -1$ does the trick, so the maximum value of $f(x)$ is $f(-1) = 1$.

76. Let x represent the edge length of the square base of the box. Because the volume of the box is 324, the box has height $324/x^2$. We minimize the cost C of materials to make the box, where

$$C = C(x) = 3x^2 + 4 \cdot x \cdot \frac{324}{x^2} = 3x^2 + \frac{1296}{x}, \qquad 0 < x < \infty.$$

Now $C'(x) = 6x - (1296/x^2)$, so $C(x) = 0$ when $x = \sqrt[3]{1296/6} = 6$. Because $C''(6) > 0$, the cost C is minimized when $x = 6$. The box we seek has a square base 6 in. on a side and height 9 in.

79. If the speed of the truck is v, then the trip time is $T = 1000/v$. So the resulting cost is

$$C(v) = \frac{10000}{v} + (1000)\left(1 + (0.0003)v^{3/2}\right),$$

so that

$$\frac{C(v)}{1000} = \frac{10}{v} + 1 + (0.0003)v^{3/2}.$$

Thus

$$\frac{C'(v)}{1000} = -\frac{10}{v^2} + \frac{3}{2}(0.0003)\sqrt{v}.$$

Then $C'(v) = 0$ when $v = (200,000/9)^{2/5} \approx 54.79$ mph. This clearly minimizes the cost, since $C''(v) > 0$ for *all* $v > 0$.

82. Let x represent the length of the internal divider. Then the field is x by $2400/x$ ft. We minimize the total length of fencing, given by:

$$f(x) = 3x + \frac{4800}{x}, \qquad 0 < x < \infty.$$

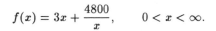

Now $f'(x) = 3 - \dfrac{4800}{x^2}$, which is zero only when $x = \sqrt{1600} = 40$. Verification: $f'(x) > 0$ if $x > 40$, and $f'(x) < 0$ if $x < 40$, so f is minimized when $x = 40$. The minimum length of fencing required for this field is 240 feet.

85. Let x represent the length of each of the dividers. Then the field is x by A/x ft. We minimize the total length of fencing, given by:

$$f(x) = (n+2)x + \frac{2A}{x}, \quad 0 < x < \infty.$$

Now $f'(x) = n+2-\dfrac{2A}{x^2}$, which is zero only when $x = \sqrt{\dfrac{2A}{n+2}}$. Verification: $f'(x) > 0$ if $x > \sqrt{\dfrac{2A}{n+2}}$ and $f'(x) < 0$ if $x < \sqrt{\dfrac{2A}{n+2}}$, so f is minimized when $x = \sqrt{\dfrac{2A}{n+2}}$. The minimum length of fencing required for this field is

$$f\left(\sqrt{\frac{2A}{n+2}}\right) = (n+2)\sqrt{\frac{2A}{n+2}} + \frac{2A\sqrt{n+2}}{\sqrt{2A}}$$
$$= \sqrt{2A(n+2)} + \sqrt{2A(n+2)}$$
$$= 2\sqrt{2A(n+2)} \text{ ft.}$$

88. Let L be the line segment in the first quadrant that is tangent to the graph of $y = 1/x$ at $(x, 1/x)$ and has endpoints $(0, c)$ and $(b, 0)$. Compute the slope of L in several ways: as the value of dy/dx at the point of tangency, as the slope of the line segment between $(x, 1/x)$ and $(b, 0)$, and as the slope of the line segment between $(x, 1/x)$ and $(0, c)$:

$$-\frac{1}{x^2} = \frac{\frac{1}{x} - 0}{x - b} \text{ so } b = 2x$$

$$-\frac{1}{x^2} = \frac{\frac{1}{x} - c}{x} : c - \frac{1}{x} = \frac{1}{x}, \text{ so } c = \frac{2}{x}.$$

Therefore the area A of the triangle is $A = A(x) = \dfrac{1}{2} \cdot 2x \cdot \dfrac{1}{x} = 1$. Because A is a constant function, every triangle has both maximal and minimal area.

91. Let x be the width of the box. Then its length is $5x$, and its height is $225/\left(5x^2\right)$. We minimize its surface area A, where

$$A(x) = 10x^2 + \frac{225}{5x^2} \cdot 2(6)x = 10x^2 + \frac{(12)(225)}{5x} \quad 0 < x < \infty.$$

Now $A'(x) = 20x - \dfrac{(12)(225)}{5x^2}$, and $A'(x) = 0$ when $x = \sqrt[3]{(6)(225)/50}$. Verification: $A'(x) > 0$ for $x > \sqrt[3]{(6)(225)/50}$, and $A'(x) < 0$ for $x < \sqrt[3]{(6)(225)/50}$, so at this critical point the surface area is minimized. The minimal surface area is 270 cm^2.

4. $\int \left(-\frac{1}{t^2}\right) dt = \frac{1}{t} + C$

10. $\int \left(2x\sqrt{x} - \frac{1}{\sqrt{x}}\right) dx = \frac{4}{5}x^{5/2} - 2x^{1/2} + C$

16. $\int (t+1)^{10} dt = \frac{1}{11}(t+1)^{11} + C$

22. $\int \frac{(3x+4)^2}{\sqrt{x}} dx = \int x^{-1/2}\left(9x^2 + 24x + 16\right) dx = \int \left(9x^{3/2} + 24x^{1/2} + 16x^{-1/2}\right) dx$

$$= \frac{18}{5}x^{5/2} + 16x^{3/2} + 32x^{1/2} + C$$

28. $\int (2\cos \pi x + 3\sin \pi x)\, dx = \frac{2}{\pi}\sin \pi x - \frac{3}{\pi}\cos \pi x + C$

31. $\frac{1}{2}\sin^2 x + C_1 = -\frac{1}{2}\cos^2 x + C_2$: $\sin^2 x + \cos^2 x + 2C_1 = 2C_2$; $1 + 2C_1 = 2C_2$; $C_2 = C_1 + \frac{1}{2}$.

34. (a) $D_x \tan x = \sec^2 x$ (b) $\int \tan^2 x\, dx = \int (\sec^2 x - 1)\, dx = \tan x - x + C.$

40. $y = \int \sqrt{x+9}\, dx = \frac{2}{3}(x+9)^{3/2} + C$; $0 = y(-4) = \frac{2}{3}(-4+9)^{3/2} + C = \frac{2}{3}\cdot 5\sqrt{5} + C$;

$y(x) = \frac{2}{3}(x+9)^{3/2} - \frac{10}{3}\sqrt{5}.$

46. $y = \int (2x+3)^{3/2}\, dx = \frac{1}{5}(2x+3)^{5/2} + C = \frac{1}{5}(2x+3)^{5/2} + \frac{257}{5}$

In the solutions for Problems 47–68, unless otherwise indicated, we will take the upward direction to be the positive direction, $s = s(t)$ for position (in feet) at time t (in seconds) with $s = 0$ corresponding to ground level, and $v(t)$ velocity at time t in ft/s, $a = a(t)$ acceleration at time t in ft/s^2. The initial position will be denoted by s_0 and the initial velocity by v_0.

49. Here it is more convenient to take the downward direction as the positive direction. Thus

$$a(t) = +32, \quad v(t) = +32t \text{ (because } v_0 = 0), \text{ and } s(t) = 16t^2 \text{ (because } s_0 = 0).$$

The stone hits the bottom when $t = 3$, and at that time we have $s = s(3) = 144$. Answer: The well is 144 feet deep.

52. First ball: $v_0 = 0$, $s_0 = 576$. So $s(t) = -16t^2 + 576$. The first ball strikes the ground at that $t > 0$ for which $s(t) = 0$; $t = 6$. Second ball: The second ball must remain aloft from time $t = 3$ until time $t = 6$, thus for 3 seconds. Reset $t = 0$ as the time it is thrown downward. Then with initial velocity v_0, the second ball has velocity and position

$$v(t) = -32t + v_0 \text{ and } s(t) = -16t^2 + v_0 t + 576$$

at time t. We require that $s(3) = 0$; that is, $0 = s(3) = -144 + 3v_0 + 576$, so that $v_0 = -144$. Answer: The second ball should be thrown straight downward with an initial velocity of 144 ft/s.

55. With $v(t) = -32t + v_0$ and $s(t) = -16t^2 + v_0 t$, we have the maximum altitude $s = 225$ occurring when $v(t) = 0$; that is, when $t = v_0/32$. So

$$225 = s(v_0/32) = (-16)(v_0/32)^2 + (v_0)^2/32 = (v_0)^2/64.$$

It follows that $v_0 = +120$. So the initial velocity of the ball was 120 ft/s.

58. Here s_0 will be the height of the building, so

$$v(t) = -32t - 25 \text{ and } s(t) = -16t^2 - 25t + s_0.$$

The velocity of impact is -153 ft/s, so we can obtain the time of impact t by solving $v(t) = -153$; $t = 4$. At this time we also have $s = 0$:

$$0 = s(4) = (-16)(16) - (25)(4) + s_0,$$

so that $s_0 = 356$. Answer: The building is 356 feet high.

61. Because $v_0 = -40$, $v(t) = -32t - 40$. Thus

$$s(t) = -16t^2 - 40t + 555.$$

Now $s(t) = -$ when $g = \frac{1}{4}\left(-5 + 2\sqrt{145}\right) \approx 4.77$ (s). The speed at impact is $|v(t)|$ for that value of t; that is, $16\sqrt{145} \approx 192.6655$ (ft/s), over 131 miles per hour.

64. Let $x(t)$ denote the distance the car has traveled t seconds after the brakes are applied; let $v(t)$ denote its velocity and $a(t)$ its acceleration at time t during the braking. Then we are given $a = -40$, so $v(t) = -40t + 88$ (because 60 mi/h is the same speed as 88 ft/s). The car comes to a stop when $v(t) = 0$; that is, when $t = 2.2$. The car travels the distance $x(2.2) \approx 96.8$. Answer: 96.8 feet.

67. (a) With the usual coordinate system, the ball has velocity $v(t) = -32t + v_0$ (ft/s) at time t (seconds) and altitude $y(t) = -16t^2 + v_0 t$ (ft). We require $y(T) = 144$ when $v(T) = 0$: $v_0 = 32T$, so

$$144 = -16t^2 + 32t = 16t^2,$$

and thus $T = 3$ and $v_0 = 96$. Answer: 96 ft/s.

(b) Now $v(t) = -\frac{26}{5}t + 96$, $y(t) = -\frac{13}{5}t^2 + 96t$. Maximum height occurs when $v(t) = 0$: $t = \frac{240}{13}$. The maximum height is

$$y\left(\frac{240}{13}\right) = -\frac{13}{5} \cdot \frac{240^2}{13^2} + 96 \cdot \frac{240}{13} = \frac{11520}{13} \approx 886 \text{ ft}.$$

Section 5.3

1. $\displaystyle\sum_{n=1}^{5} n^2$.

4. $\displaystyle\sum_{i=1}^{5} \frac{1}{i^2}$.

7. $\displaystyle\sum_{n=1}^{5} \left(\frac{2}{3}\right)^n$.

10. -32

16. 1330

19. $\displaystyle\lim_{n\to\infty} \frac{1^2 + 2^2 + \cdots + n^2}{n^3} = \lim_{n\to\infty} \sum_{i=1}^{n} \left(\frac{i}{n}\right)^2 \cdot \frac{1}{n} = \int_0^1 x^2\, dx = \frac{1}{3}$.

22. $\displaystyle\sum_{i=1}^{n} (2i-1)^2 = \sum_{i=1}^{n} 4i^2 - 4i + 1 = 4(\frac{1}{3}n^3 + \frac{1}{2}n^2 + \frac{1}{6}n) - 4(\frac{1}{2}n^2 + \frac{1}{2}n) + n = \frac{1}{3}(4n^3 - n)$.

25. $\displaystyle \underline{A}_6 = \sum_{i=1}^{6} \left[2\left(\frac{i-1}{2}\right) + 3\right] \cdot \frac{1}{6}$, $\qquad \overline{A}_6 = \sum_{i=1}^{6} \left[2 \cdot \frac{i}{2} + 3\right] \cdot \frac{1}{6}$.

28. $\displaystyle \underline{A}_5 = \sum_{i=1}^{5} \left(\frac{2i+3}{5}\right)^2 \cdot \frac{1}{5}$, $\qquad \overline{A}_5 = \sum_{i=1}^{5} \left(\frac{2i+4}{5}\right)^2 \cdot \frac{1}{5}$.

31. $A_{10} = \displaystyle\sum_{i=1}^{10} \left(\dfrac{i-1}{10}\right)^3 \cdot \dfrac{1}{10}, \qquad \overline{A}_{10} = \displaystyle\sum_{i=1}^{10} \left(\dfrac{i}{10}\right)^3 \cdot \dfrac{1}{10}.$

37. $\displaystyle\sum_{i=1}^{n} \left(\dfrac{3i}{n}\right)^3 \cdot \dfrac{3}{n} = \dfrac{81n^2(n+1)^2}{4n^4} \to \dfrac{81}{4}$ as $n \to \infty.$

40. $\displaystyle\sum_{i=1}^{n} \left(9 - \left(\dfrac{3i}{n}\right)^2\right)^2 \cdot \dfrac{3}{n} = 9n \cdot \dfrac{3}{n} - \dfrac{27n(n+1)(2n+1)}{6n^3} \to 27 - 9 = 18$ as $n \to \infty.$

Section 5.4

1. $\displaystyle\int_1^3 (2x-1)\, dx$

4. $\displaystyle\int_0^3 (x^3 - 3x^2 + 1)\, dx$

7. $\displaystyle\int_3^8 \dfrac{1}{\sqrt{1+x}}\, dx$

10. $\displaystyle\int_0^{\pi/4} \tan x\, dx$

13. $\displaystyle\sum_{i=1}^{n} f(x_i^*)\, \Delta x = \sum_{i=1}^{5} \dfrac{1}{1+(i)} = 1.45.$

16. $\displaystyle\sum_{i=1}^{n} f(x_i^*)\, \Delta x = \sum_{i=1}^{6} \left((1+i/2)^2 + 2(1+i/2)\right)(1/2) = 41.375$

19. $\displaystyle\sum_{i=1}^{6} \left(\cos \dfrac{i\pi}{6}\right) \cdot \dfrac{\pi}{6} = -\pi/6.$

22. 0.16

28. 4.092967 (rounded)

31. $33/100$

34. 7.505140

40. $\displaystyle\sum_{i=1}^{6} \left(\sin \pi \cdot \dfrac{2i-1}{12}\right) \cdot \dfrac{1}{6} = \dfrac{\sqrt{2} + \sqrt{6}}{6}.$

43. $\displaystyle\sum_{i=1}^{n} \left(\dfrac{2i}{n}\right)^2 \cdot \dfrac{2}{n} = \dfrac{8n(n+1)(2n+1)}{6n^3} \to \dfrac{8}{3}$ as $n \to \infty.$

46. $\displaystyle\sum_{i=1}^{n} \left(4 - 3\left(1 + \dfrac{4i}{n}\right)\right) \cdot \dfrac{4}{n} = \dfrac{16}{n} \cdot n - \dfrac{12}{n} \cdot n - \dfrac{48n(n+1)}{2n^2} \to 4 - 24 = -20$ as $n \to \infty.$

49. Take $x_i = \dfrac{bi}{n}$ and $\Delta x = \dfrac{b}{n}$. The integral is equal to $\displaystyle\lim_{n\to\infty} \dfrac{n(n+1)}{2n^2} b^2 = \dfrac{1}{2}b^2$

Section 5.5

1. $\displaystyle\int_0^1 (3x^2 + 2\sqrt{x} + 3\sqrt[3]{x})\, dx = \left[x^3 + \tfrac{4}{3}x^{3/2} + \tfrac{9}{4}x^{4/3}\right]_0^1 = \tfrac{55}{12}.$

4. $\displaystyle\int_{-2}^{-1} \dfrac{1}{x^4}\, dx = \left[-\dfrac{1}{3x^3}\right]_{-2}^{-1} = \dfrac{7}{24}.$

7. $\displaystyle\int_{-1}^{0} (x+1)^3\, dx = \left[\tfrac{1}{4}(x+1)^4\right]_{-1}^{0} = \tfrac{1}{4}.$

10. $\int_1^4 \frac{1}{\sqrt{x}}\, dx = \left[2\sqrt{x} \right]_1^4 = 2\sqrt{4} - 2\sqrt{1} = 2.$

13. $\int_{-1}^1 x^{99}\, dx = \left[\frac{1}{100} x^{100} \right]_{-1}^1 = 0.$

16. $\int_1^2 (x^2 + 1)^3\, dx = \int_1^2 (x^6 + 3x^4 + 3x^2 + 1)\, dx = \left[\frac{1}{7}x^7 + \frac{3}{5}x^5 + x^3 + x \right]_1^2 = \frac{1566}{35} \approx 44.742857.$

19. $\int_1^8 x^{2/3}\, dx = \left[\frac{3}{5}x^{5/3} \right]_1^8 = \frac{93}{5}.$

22. $\int_0^4 \sqrt{3t}\, dt = \left[\frac{2}{3}t^{3/2}\sqrt{3} \right]_0^4 = \frac{14}{3}\sqrt{3}.$

25. $\int_1^4 \frac{x^2 - 1}{\sqrt{x}}\, dx = \int_1^4 (x^{3/2} - x^{-1/2})\, dx = \left[\frac{2}{5}x^{5/2} - 2x^{1/2} \right]_1^4 = \frac{52}{5}.$

28. $\int_0^{\pi/2} \cos 2x\, dx = \left[\frac{1}{2}\sin 2x \right]_0^{\pi/2} = 0.$

31. $\int_0^\pi \sin 5x\, dx = \left[-\frac{1}{5}\cos x \right]_0^\pi = \frac{2}{5}.$

34. $\int_0^5 \sin \frac{\pi x}{10}\, dx = \left[-\frac{10}{\pi}\cos \frac{\pi x}{10} \right]_0^5 = \frac{10}{\pi}.$

37. Choose $x_i = i/n$, $\Delta x = 1/n$, $x_0 = 0$, and $x_n = 1$. Then the limit in question is the limit of a Riemann sum for the function $f(x) = 2x - 1$ on the interval $0 \le x \le 1$, and its value is therefore

$$\int_0^1 (2x - 1)\, dx = \left[x^2 - x \right]_0^1 = 1 - 1 = 0.$$

40. This limit is the integral of $f(x) = x^3$ on the interval $0 \le x \le 1$, and it is therefore equal to $\frac{1}{4}$.

43. This integral is $25\pi/4$ because it is the area of a quarter-circle of radius 5; specifically, the first-quadrant portion of such a circle with center at the origin.

49.
$$\frac{0.2}{1.2} + \frac{0.2}{1.4} + \frac{0.2}{1.6} + \frac{0.2}{1.8} + \frac{0.2}{2.0} \le \int_1^2 \frac{1}{x}\, dx \le \frac{0.2}{1.0} + \frac{0.2}{1.2} + \frac{0.2}{1.4} + \frac{0.2}{1.6} + \frac{0.2}{1.8}.$$

Therefore
$$\frac{1627}{2520} \le \int_1^2 \frac{1}{x}\, dx \le \frac{1879}{2520}.$$

In decimal form—rounded down on the left, up on the right—

$$0.6456349 \le \int_1^2 \frac{1}{x}\, dx \le 0.7456350.$$

Section 5.6

1. $\frac{1}{2}\int_0^2 x^4\, dx = 16/5.$

4. $\frac{1}{4}\int_0^4 8x\, dx = 16.$

10. $\frac{2}{\pi}\int_0^{\pi/2} \sin 2x\, dx = \frac{2}{\pi}.$

16. $\int_{-1}^1 (x^3 + 2)^2\, dx = \left[\frac{1}{7}x^7 + x^4 + 4x \right]_{-1}^1 = \frac{58}{7}.$

22. On the interval $-1 \leq x \leq 1$ of integration, the integrand is equal to $x-2$, so the value of the integral is -4.

25. Split the integral into three integrals—one on the interval from -2 to -1, one on the interval from -1 to 1, and one on the interval from 1 to 2.

28. $\displaystyle\int_5^{10} \frac{dx}{\sqrt{x-1}} = \left[2\sqrt{x-1}\right]_5^{10} = 2.$

31. $\displaystyle\int_{-3}^0 (x^3 - 9x)\,dx - \int_0^3 (x^3 - 9x)\,dx = \left[\tfrac{1}{4}x^4 - \tfrac{9}{2}x^2\right]_{-3}^0 - \left[\tfrac{1}{4}x^4 - \tfrac{9}{2}x^2\right]_0^3 = \tfrac{81}{4} + \tfrac{81}{4} = \tfrac{81}{2}.$

34. $P_{AV} = \dfrac{1}{10}\displaystyle\int_0^{10} \left(100 + 10t + (0.5)t^2 + (0.02)t^3\right)dt = \dfrac{1}{10}\left[100t + 5t^2 + \tfrac{1}{6}t^3 + (0.005)t^4\right]_0^{10}$

$= \tfrac{1}{10}\left(1000 + 500 + \tfrac{1000}{6} + 50\right) = \tfrac{515}{3}$, approximately 171.667.

37. $T_{AV} = \dfrac{1}{10}\displaystyle\int_0^{10} (40x - 4x^2)\,dx = \dfrac{1}{10}\left[20x^2 - \tfrac{4}{3}x^3\right]_0^{10} = \dfrac{1}{10}\left(2000 - \dfrac{4000}{3}\right) = \dfrac{200}{3}.$

40. Let $v(t)$ represent velocity of the car. Then $v(t) = at$ and the position of the car at time t is $x(t) = \tfrac{1}{2}at^2$. So the final velocity, at time $t = T$, is $v(T) = aT$. Average velocity during the T seconds:

$$v_{AV} = \frac{1}{T}\int_0^T at\,dt = \frac{1}{T}\left[\frac{1}{2}at^2\right]_0^T = \frac{1}{2}aT.$$

The final position, at time $t = T$, is $x(T) = \tfrac{1}{2}aT^2$. Average position during the T seconds:

$$x_{AV} = \frac{1}{T}\int_0^T \frac{1}{2}at^2\,dt = \frac{1}{T}\cdot\frac{a}{2}\cdot\left[\frac{1}{3}t^3\right]_0^T = \frac{a}{6T}\cdot T^3 = \frac{1}{6}aT^2.$$

43. $h'(z) = (z-1)^{1/3}$

46. $G'(x) = \dfrac{x}{x^2 + 1}$

49. $G'(x) = (x^3 + 1)^{1/2}$

52. $f'(x) = (1 - \sin^2 x)^{1/2}\cos x = |\cos x|\cos x.$

58. $dy = \dfrac{1}{1+x^2}\,dx$: $y(x) = \dfrac{\pi}{4} + \displaystyle\int_1^x \dfrac{1}{1+t^2}\,dt.$

Section 5.7

4. $\displaystyle\int \sqrt{2x+1}\,dx = \tfrac{1}{3}(2x+1)^{3/2} + C$

10. $\displaystyle\int \csc^2 5x\,dx = -\tfrac{1}{5}\cot 5x + C$

16. $\displaystyle\int \frac{x^2}{\sqrt[3]{x^3 + 1}}\,dx = \tfrac{1}{2}\left(x^3 + 1\right)^{2/3} + C$

22. $\displaystyle\int \sin^5 3z \cos 3z\,dz = \tfrac{1}{18}\sin^6 3z + C$

28. $\displaystyle\int \frac{(x+2)\,dx}{(x^2 + 4x + 3)^2} = -\tfrac{1}{4}(x^2 + 4x + 3)^{-2} + C$

34. $\displaystyle\int_1^4 \frac{(1 + \sqrt{x})^4}{\sqrt{x}}\,dx = \left[\frac{2}{5}(1 + \sqrt{x})^5\right]_1^4 = \frac{422}{5}$

40. $\displaystyle\int_0^{\pi/2} \sec^2 \frac{x}{2}\,dx = \left[2\tan\frac{x}{2}\right]_0^{\pi/2} = 2$

46. $\int \cos^2 x \, dx = \int \frac{1}{2}(1 + \cos 2x) \, dx = \frac{1}{2}(x + \frac{1}{2}\sin 2x) + C$

52. $\int_0^{\pi/2} \cos^3 x \, dx = \int_0^{\pi/2} \cos x(1 - \sin^2 x) \, dx = \int_0^{\pi/2} (\cos x - \sin^2 x \cos x) \, dx$

$$= \left[\sin x - \frac{1}{3}\sin^3 x\right]_0^{\pi/2} = 1 - \frac{1}{3} = \frac{2}{3}.$$

58. $\int_{-a}^a f(x) \, dx = \int_{-a}^0 f(x) \, dx + \int_0^a f(x) \, dx = -\int_a^0 f(-u) \, du + \int_0^a f(x) \, dx$

$$= \int_0^a f(u) \, du + \int_0^a f(x) \, dx = 2\int_0^a f(x) \, dx.$$

Section 5.8

1. $A = \int_1^3 \frac{1}{x^2} \, dx = \left[-\frac{1}{x}\right]_1^3 = 1 - \frac{1}{3} = \frac{2}{3}.$

4. $A = \int_0^1 \frac{4x}{(x^2+1)^{3/2}} \, dx = \left[-4\left(x^2+1\right)^{-1/2}\right]_0^1 = 4\left(1 - \frac{1}{\sqrt{2}}\right).$

7. $A = \int_0^{\pi/2} \sin^3 x \cos x \, dx = \left[\frac{1}{4}\sin^4 x\right]_0^{\pi/2} = \frac{1}{4}.$

10. $A = \int_0^2 \left(x - (x^2 - x)\right) \, dx = \left[x^2 - \frac{1}{3}x^3\right]_0^2 = \frac{4}{3}.$

16. $A = \int_0^4 \left(4x - x^2\right) \, dx = \left[2x^2 - \frac{1}{3}x^3\right]_0^4 = \frac{32}{3}.$

22. $A = \int_{-2}^2 \left(4 - x^2\right) \, dx = \left[4x - \frac{1}{3}x^3\right]_{-2}^2 = \frac{32}{3}.$

28. $A = \int_0^1 \left(\sqrt{x} - x^2\right) \, dx = \left[\frac{2}{3}x^{3/2} - \frac{1}{3}x^3\right]_0^1 = \frac{1}{3}.$

34. $\dfrac{32}{27}$

40. $4\sqrt{6}$

46. Let $A_b = \int_1^b \frac{1}{x^2} \, dx$. Then $A_b = \left[-\frac{1}{x}\right]_1^b = 1 - \frac{1}{b}$. Therefore $A = \lim_{b \to \infty} A_b = 1$.

49. Given:

$$A(u) = u^5 = \int_0^u f(x) \, dx.$$

We differentiate with respect to u to find that $A'(u) = 5u^4 = f(u)$, and therefore $f(x) = 5x^4$.

Section 5.9

1. $T_4 = 8$, and the true value of the integral is also 8.

4. $T_4 = 0.71$, and the true value of the integral is $\frac{2}{3}$.

7. $M_4 = 8.$

10. $M_4 = 0.65.$

13. $T_4 = 8.75$, $S_4 \approx 8.6667$, integral: $\frac{26}{3}$.

16. $T_4 \approx 1.2182$, $S_4 \approx 1.2189$, integral: $\frac{2}{3}(2\sqrt{2} - 1) \approx 1.218951417.$

19. $T_8 \approx 8.5499$, $S_8 \approx 8.5509$, integral: approximately 8.550733044.

22. **(a)** 28.3 **(b)** 28.6

28. $f^{(4)}(x) = \dfrac{24}{x^5}$, and the maximum of $\left|f^{(4)}(x)\right|$ for $1 \leq x \leq 2$ is 24. We take $M = 24$, $a = 1$, $b = 2$, and ask for n large enough that
$$\frac{M(b-a)^5}{180n^4} < 0.000005;$$
thus $n^4 > \dfrac{2}{(15)(0.000005)}$, and therefore $n \geq 14$.

Chapter 5 Miscellaneous

1. $\displaystyle\int \frac{x^5 - 2x + 5}{x^3}\, dx = \tfrac{1}{3}x^3 + \frac{2}{x} - \frac{5}{2x^2} + C$

4. $\displaystyle\int 7(2x+3)^{-3}\, dx = -\tfrac{7}{4}(2x+3)^{-2} + C$

10. $\displaystyle\int 3x\,(1+3x^2)^{-1/2}\, dx = \sqrt{1+3x^2} + C$

16. $\displaystyle\int (1+x^{1/2})^{-2}\,x^{-1/2}\, dx = \int (1+u)^{-2}\,2\,du = -\frac{2}{1+u} + C = -\frac{2}{1+\sqrt{x}} + C$

22. $\displaystyle\int \frac{x + 2x^3}{(x^4 + x^2)^3}\, dx = \int \tfrac{1}{2}u^{-3}\, du = -\tfrac{1}{4}\left(x^4 + x^2\right)^{-2} + C$

28. $y(x) = \displaystyle\int \frac{2}{\sqrt{x+5}}\, dx = 4(x+5)^{1/2} + C$; $y(4) = 3 = 12 + C$, so $C = -9$; $y(x) = 4\sqrt{x+5} - 9$.

34. With the usual notation, $a = +8$, $v = 8t$, and $s = 4t^2$. When $v = 88$, $t = 11$, so the distance traveled then will be $(4)(11)^2 = 484$ feet.

37. Assume that the brakes are applied at time $t = 0$ (seconds), that $s(t)$ is the distance the car subsequently travels, that $v(t)$ is its velocity at time t, and that its constant deceleration is $-a$ (where $a > 0$). Then
$$s = -\frac{1}{2}at^2 + v_0 t \quad \text{and} \quad v = -at + v_0.$$

Now $v = 0$ when $t = v_0/a$, and it follows from the first equation above that $a = 22$. So $s = -11t^2 + v_0 t$ and $v = -22t + v_0$. But $v = 0$ when $t = 4$, and at this time we have $s = 176$ feet. The point of this problem is that *doubling* the speed *quadruples* the stopping distance.

40. $\displaystyle\sum_{k=1}^{100}\left(\frac{1}{k} - \frac{1}{k+1}\right) = 1 - \frac{1}{2} + \frac{1}{2} - \frac{1}{3} + \frac{1}{3} - \frac{1}{4} + \cdots + \frac{1}{98} - \frac{1}{99} + \frac{1}{99} - \frac{1}{100} + \frac{1}{100} - \frac{1}{101} = 1 - \frac{1}{101} = \frac{100}{101}.$

43. $\displaystyle\lim_{n\to\infty}\sum_{i=1}^{n}\frac{\Delta x}{\sqrt{x_i^*}} = \int_1^2 x^{-1/2}\, dx = \left[2\sqrt{x}\,\right]_1^2 = 2\left(\sqrt{2} - 1\right).$

46. $\displaystyle\lim_{n\to\infty}\frac{1^{10} + 2^{10} + 3^{10} + \cdots + n^{10}}{n^{11}} = \int_0^1 x^{10}\, dx = \left[\frac{1}{11}x^{11}\right]_0^1 = \frac{1}{11}.$

52. $\displaystyle\int \frac{(1-\sqrt[3]{x})^2}{\sqrt{x}}\, dx = \int \left(x^{-1/2} - 2x^{-1/6} + x^{1/6}\right)\, dx = 2x^{1/2} - \tfrac{12}{5}x^{5/6} + \tfrac{6}{7}x^{7/6} + C.$

55. You may use the substitution $u = x^{3/2}$, $du = \tfrac{3}{2}x^{1/2}$.

58. $\displaystyle\int_1^2 \frac{2t+1}{\sqrt{t^2+t}}\, dt = \left[2(t^2+t)^{1/2}\right]_1^2 = 2\sqrt{2}\left(\sqrt{3} - 1\right)$ (using the substitution $u = t^2 + t$).

61. One easy substitution is $u = 1 + \sqrt{t}$.

64. $A = \displaystyle\int_{-1}^{1}(1 - x^3)\, dx = \left[x - \frac{1}{4}x^4\right]_{-1}^{1} = 2.$

67. The curves cross at $(-1, 1)$ and at $(1, 1)$; we obtain the total area by doubling the integral of $2-x^2-x^4$ over the interval $0 \le x \le 1$.

70. The curves cross at $(-1, 1)$ and at $(1, 1)$. By symmetry,

$$A = 2 \int_0^1 \left(2 - x^2 - x^{2/3}\right) dx = 2 \left[2x - \frac{1}{3}x^3 - \frac{3}{5}x^{5/3} \right]_0^1 = \frac{32}{15}.$$

73. Differentiation of both sides of the given equation yields

$$2x = \left(1 + [f(x)]^2\right)^{1/2}, \quad \text{so that } 4x^2 = 1 + [f(x)]^2, \quad \text{and therefore } f(x) = \left(4x^2 - 1\right)^{1/2}.$$

This answer must be tested by substitution in the original equation because it was obtained under the assumption that such a function actually exists.

76. $T_6 \approx 2.812254$ and $S_6 \approx 2.828502$. For the exact value:

$$(1 - \cos x)^{1/2} = \sqrt{2} \left(\frac{1 - \cos x}{2}\right)^{1/2} = \sqrt{2} \left|\sin \frac{x}{2}\right|.$$

Therefore

$$\int_0^\pi (1 - \cos x)^{1/2} dx = \int_0^\pi \sqrt{2} \left(\frac{1 - \cos x}{2}\right)^{1/2} dx = \sqrt{2} \int_0^\pi \left|\sin \frac{x}{2}\right| dx = \left[-2\sqrt{2} \cos \frac{x}{2}\right]_0^\pi = 2\sqrt{2}.$$

1. $\displaystyle\lim_{n\to\infty}\sum_{i=1}^{n} 2x_i^{*}\,\Delta x = \int_0^1 2x\,dx = 1.$

4. $\displaystyle\lim_{n\to\infty}\sum_{i=1}^{n}\left(3\left(x_i^{*}\right)-1\right)\Delta x = \int_{-1}^{3}\left(3x^2-1\right)dx = 24.$

7. $\displaystyle\lim_{n\to\infty}\sum_{i=1}^{n}\left(2m_i-1\right)\Delta x = \int_{-1}^{3}\left(2x-1\right)dx = 4.$

10. $\displaystyle\lim_{n\to\infty}\sum_{i=1}^{n} m_i\cos\left(m_i\right)^2\Delta x = \int_0^{\sqrt{\pi}} x\cos\left(x^2\right)dx = \left[\frac{1}{2}\sin\left(x^2\right)\right]_0^{\sqrt{\pi}} = 0.$

13. $\displaystyle\lim_{n\to\infty}\sum_{i=1}^{n}\sqrt{1+\left[f\left(x_i^{*}\right)\right]^2}\,\Delta x = \int_0^{10}\left(1+[f(x)]^2\right)^{1/2}dx$

16. $\displaystyle M = \int_0^{25}(60-2x)\,dx\,dx = \left[60x-x^2\right]_0^{25} = 875$

19. $\displaystyle\int_0^{10} -32t\,dt = -320$ is the net distance; $\displaystyle\int_0^{10} 32t\,dt = 320$ is the total distance.

22. $\displaystyle\int_0^5 |2t-5|\,dt = 2\int_0^{5/2}(5-2t)\,dt = \frac{25}{2}$ is the net and the total distance.

28. Net distance: 2. Total distance: $2\sqrt{2}$.

31. $\displaystyle M = \int_0^5 2\pi x\left(25-x^2\right)dx = \left[25\pi x^2-\frac{1}{2}\pi x^4\right]_0^5 = \frac{625}{2}\pi.$

34. $\displaystyle\int_0^{20}(13+t)\,dt = 460$ (thousands of births).

37. $r(0) = 0.1 = a - b$, $r(182.5) = a + b$. So $a = 0.3$ and $b = 0.2$. Therefore the average total annual rainfall will be
$$R = \int_0^{365}\left(0.3-0.2\cos\frac{2\pi t}{365}\right)dt = (0.3)(365) - \frac{36.5}{\pi}\sin(2\pi) = 109.5 \text{ (inches)}.$$

Section 6.2

1. $\displaystyle V = \int_0^1 \pi x^4\,dx = \frac{\pi}{5}.$

4. $\displaystyle V = \int_{0.1}^1 \frac{\pi}{x^2}\,dx = 9\pi.$

7. $\displaystyle V = \int_0^1\left(\pi x-\pi x^4\right)dx = \frac{3}{10}\pi.$

10. $\displaystyle V = \int_{-2}^3 \pi\left[(y+6)^2-(y^2)^2\right]dy = \frac{500}{3}\pi.$

13. $\displaystyle V = \int_0^1 \pi\left(\sqrt{1-y}\right)^2 dy = \frac{\pi}{2}.$

16. $\displaystyle V = \int_0^1 \pi\left[\left(2+\sqrt{1-y}\right)^2-\left(2-\sqrt{1-y}\right)^2\right]dy = \frac{16}{3}\pi.$

19. $\displaystyle V = \int_0^4 \pi\left(\sqrt{y}\right)^2 dy = 8\pi.$

22. $V = 2 \displaystyle\int_0^2 \pi \left[(9 - x^2)^2 - (x^2 + 1)^2 \right] dx = \dfrac{640}{3} \pi \approx 670.206433.$

Note how we used the fact that the plane region is symmetric about the y-axis.

25. $V = \displaystyle\int_0^6 \pi x \, dx - \int_3^6 2\pi(x - 3) \, dx = 9\pi.$

28. $V_b = \displaystyle\int_1^b \dfrac{\pi}{x^4} \, dx = \dfrac{\pi}{3} \left(1 - \dfrac{1}{b^3} \right).$ Therefore $V = \displaystyle\lim_{b \to \infty} V_b = \dfrac{\pi}{3}.$

31. $V = 2 \displaystyle\int_0^a \sqrt{3} \, (a^2 - x^2) \, dx = \dfrac{4}{3} a^3 \sqrt{3}.$

34. Let e denote the length of each edge of the base of the pyramid. Imagine a cross section at level y—height y above the base. Then (by similar triangles)

$$\frac{h - y}{x} = \frac{h}{e},$$

so that the cross-sectional area is $A_y = \left(\dfrac{h - y}{h} \right)^2 A.$ Thus

$$V = \int_0^h \left(1 - \frac{y}{h} \right)^2 A \, dy = \frac{1}{3} Ah.$$

37. See the figure below, which shows the nearest quarter of each of the two cylinders. This implies that the volume of the intersection illustrated is one-eighth of the total volume of their intersection. Note that we have chosen the x-axis and the z-axis as the centerlines of the cylinders; the y-axis is vertical.

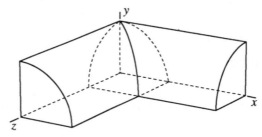

40. $V \approx \frac{25}{3} \left[\pi(60)^2 + 4\pi(55)^2 + 2\pi(50)^2 + 4\pi(35)^2 + 0 \right] = \frac{25}{3} \pi(25,600) \approx 670,200$ (cubic feet).

Section 6.3

1. $V = \displaystyle\int_0^2 2\pi x \, (x^2) \, dx = 8\pi.$

4. $V = \displaystyle\int_0^2 2\pi x \, (8 - 2x^2) \, dx = 16\pi.$

7. $V = \displaystyle\int_0^1 2\pi y \, (3 - 3y) \, dy = \pi.$

10. $V = \displaystyle\int_0^3 2\pi x \, (3x - x^2) \, dx = \dfrac{27\pi}{2}.$

13. $V = \displaystyle\int_0^1 2\pi x \, (x - x^3) \, dx = \dfrac{4\pi}{15}.$

16. $V = \displaystyle\int_0^2 2\pi x \, (x^3) \, dx = \dfrac{64\pi}{5}.$

19. $V = \displaystyle\int_{-1}^{1} 2\pi(2 - x)\left(x^2\right) dx = \dfrac{8\pi}{3}.$ This is one of many examples in which you do *not* obtain the correct answer by doubling the value of the integral from 0 to 1.

22. $V = \displaystyle\int_{0}^{1} 2\pi(2 - y)\left(\sqrt{y} - y\right) dy = \dfrac{8\pi}{15}.$

25. $V = \displaystyle\int_{-1}^{1} 2\pi(1 - y)\left(2 - 2y^2\right) dy = \dfrac{16\pi}{3}.$

28. $V = \displaystyle\int_{0}^{1} 2\pi(y + 1)\left(\sqrt{y} - y^2\right) dy = \dfrac{29\pi}{30}.$

31. $V = 2\displaystyle\int_{0}^{a} 2\pi x \dfrac{b}{a}\left(a^2 - x^2\right)^{1/2} dx = \dfrac{4}{3}\pi a^2 b.$

34. (a) $V = \displaystyle\int_{-1}^{2} 2\pi(x + 2)\left(x + 2 - x^2\right) dx = \dfrac{45\pi}{2}.$

    (b) $V = \displaystyle\int_{-1}^{2} 2\pi(3 - x)\left(x + 2 - x^2\right) dx = \dfrac{45\pi}{2}.$

37. $a^2 + \tfrac{1}{4}h^2 = r^2$, so $r^2 - a^2 = \tfrac{1}{4}h^2$. Therefore $V = \tfrac{4}{3}\pi\left(\tfrac{1}{4}h^2\right)^{3/2}$. So the answer to part (a) is $V = \tfrac{1}{6}\pi h^3$.

    (b) The answer involves neither the radius r of the sphere nor the radius a of the hole; it depends only upon the length of the hole.

Section 6.4

1. $\displaystyle\int_{0}^{1} \sqrt{1 + 4x^2}\, dx$

4. $\displaystyle\int_{-1}^{1} \sqrt{1 + \tfrac{16}{9}x^{2/3}}\, dx$

7. $\displaystyle\int_{-1}^{2} \sqrt{1 + 16y^6}\, dy$

10. $\displaystyle\int_{0}^{2} \dfrac{2}{\sqrt{4 - x^2}}\, dx$

13. $\displaystyle\int_{0}^{1} 2\pi\left(x - x^2\right)\sqrt{4x^2 - 4x + 2}\, dx$

16. $\displaystyle\int_{0}^{1} 2\pi\left(x - x^3\right)\sqrt{9x^4 - 6x^2 + 2}\, dx$

19. $\displaystyle\int_{1}^{4} 2\pi(x + 1)\sqrt{1 + \tfrac{9}{4}x}\, dx$

22. $\dfrac{dx}{dy} = \sqrt{y - 1}$, so $L = \displaystyle\int_{1}^{5} \left(1 + \left(\sqrt{y - 1}\right)^2\right)^{1/2} dy = \tfrac{2}{3}\left(5\sqrt{5} - 1\right).$

25. $\dfrac{dy}{dx} = \dfrac{6x^4 - 8y}{4x}$. But $8y = \dfrac{2x^6 + 1}{x^2}$, so $\dfrac{dy}{dx} = x^3 - \tfrac{1}{4}x^{-3}$. Therefore $1 + \left(\dfrac{dy}{dx}\right)^2 = \left(x^3 + \tfrac{1}{4}x^{-3}\right)^2$, so

$$L = \int_{1}^{2} \left(x^3 + \tfrac{1}{4}x^{-3}\right) dx = \tfrac{123}{32} = 3.84375.$$

28. $2(y - 3)\dfrac{dy}{dx} = 12(x + 2)^2$, so $\dfrac{dy}{dx} = \dfrac{6(x + 2)^2}{y - 3}$. Thus

$$1 + \left(\dfrac{dy}{dx}\right)^2 = 1 + \dfrac{36(x + 2)^2}{(y - 3)^2} = 1 + \dfrac{36(x + 2)^4}{4(x + 2)^3} = 1 + 9(x + 2) = 9x + 19.$$

Therefore
$$L = \int_{-1}^{2} (9x + 19)^{1/2}\, dx = \frac{2}{27}\left(37\sqrt{37} - 10\sqrt{10}\right).$$

31. $1 + \left(\dfrac{dy}{dx}\right)^2 = \left(x^4 + \frac{1}{4}x^{-4}\right)^2$. Therefore $A = \displaystyle\int_{1}^{2} 2\pi x\left(x^4 + \frac{1}{4}x^{-4}\right)\, dx = \frac{339}{16}\pi$.

34. $A = \displaystyle\int_{1}^{2} 2\pi x\sqrt{1 + x}\, dx$. Let $u = 1 + x$; then $x = u - 1$ and $dx = du$. As x takes on values from 1 to 2, u takes on values from 2 to 3. So
$$A = \int_{2}^{3} 2\pi(u - 1)u^{1/2}\, du = \frac{8}{15}\pi\left(6\sqrt{3} - \sqrt{2}\right).$$

40. Take $y = f(x) = (r^2 - x^2)^{1/2}$, $-r \le x \le r$. Then $\dfrac{dy}{dx} = -\dfrac{x}{\sqrt{r^r - x^2}}$. So
$$A = \int_{-r}^{r} 2\pi\sqrt{r^2 - x^2}\ \sqrt{1 + \frac{x^2}{r^2 - x^2}}\, dx = 2\pi\int_{-r}^{r} \left(r^2\right)^{1/2}\, dx = 4\pi r^2.$$

Section 6.5

1. $y^{-1/2} = 2x\, dx$; $2y^{1/2} = x^2 + C$; $y(x) = \frac{1}{4}\left(x^2 + C\right)^2$.

4. $y^{-3/2}\, dy = x^{3/2}\, dx$; $-2y^{-1/2} = \frac{2}{5}x^{5/2} + K$; $y^{-1/2} = \frac{1}{5}\left(C - x^{5/2}\right)$; $y(x) = 25\left(C - x^{5/2}\right)^{-2}$.

7. $\left(1 + y^{1/2}\right)dy = \left(1 + x^{1/2}\right)dx$; $y + \frac{2}{3}y^{3/2} = x + \frac{2}{3}x^{3/2} + K$; $3y + 2y^{3/2} = 3x + 2x^{3/2} + C$.

10. $2y + \frac{3}{2}y^{-2} = x + x^{-1} + C$

13. $4y^3\, dy = 1\, dx$; $y^4 = x + C$. $1 = 0 + C$, so $C = 1$. $y^4 = x + 1$.

16. $y\, dy = x\, dx$; $\frac{1}{2}y^2 = \frac{1}{2}x^2 + K$; $y^2 = x^2 + C$. $25 = 9 + C$, so $C = 16$. $y^2 = x^2 + 16$; that is, $y(x) = \sqrt{x^2 + 16}$. (We take the positive sign at the last step because $y(3) > 0$.)

19. $y^{-2}\, dy = (3x^2 - 1)\, dx$; $-\left(y^{-1}\right) = x^3 - x + K$; $y(x) = \dfrac{1}{C + x - x^3}$. $1 = \dfrac{1}{C}$: $y(x) = \dfrac{1}{1 + x - x^3}$.

22. If $P(t) = \left(\frac{1}{2}kt P_0^{1/2}\right)^2$, then
$$\frac{dP}{dt} = 2\left(\frac{1}{2}kt + P_0^{1/2}\right)\left(\frac{1}{2}k\right) = k\left(\frac{1}{2}kt + P_0^{1/2}\right) = kP^{1/2} \quad\text{and}\quad P(0) = \left(P_0^{1/2}\right)^2 = P_0.$$

Thus we have verified that the proposed solution satisfies both the differential equation and the associated initial condition.

25. $P(t) = \dfrac{2}{1 - 2kt}$; $4 = P(3) = \dfrac{2}{1 - 6k}$, so $k = \frac{1}{12}$. Now $P(6)$ is undefined, but $P(t) \to +\infty$ as $t \to 6$ from below. So the rabbit population explodes in the second three-month period.

28. In the notation of Section 6.5 of the text, we have $c = 1$, $g = 32$, $a = \pi/144$, and $A(y) = 9\pi$. So Eq. (20) takes the form
$$9\pi\frac{dy}{dt} = -\frac{\pi}{144}\sqrt{64y}, \quad\text{and hence}\quad y^{-1/2}\frac{dy}{dt} = -\frac{1}{162}.$$

It follows that $y^{1/2} = C - (t/324)$. Therefore
$$y(t) = \left(C - \frac{t}{324}\right)^2.$$

Now $9 = y(0) = C^2$, so $C = 3$ (not -3, else y would never be zero). Thus

$$y(t) = \left(3 - \frac{t}{324}\right)^2$$

where the units are feet and seconds (because those are the units that we used for g and other assorted constants and variables in the problem). The tank will be empty when $y = 0$; that is, when $t = 972$ (s). Hence the tank will require approximately 16 min 12 sec to empty.

31. Let $V(y)$ denote the volume of water in the tank when the water has depth y. Then $\dfrac{dV}{dt} = -C_0\sqrt{y}$, and $V(y) = \displaystyle\int_0^y \pi u^{3/2}\, du$. Therefore

$$\pi y^{3/2} \frac{dy}{dt} = -C_0 y^{1/2}$$
$$\pi y\, dy = -C_0\, dt$$
$$\frac{1}{2}\pi y^2 = -C_0 t + K_0$$
$$y^2 = -Kt + C.$$

Let $t = 0$ at noon. Then $y(0) = 12$, so $C = 144$. Since $y(1) = 6$, $36 = 144 - K$, so $K = 108$. Therefore $y^2 = 144 - 108t$. When $y = 0$, $t = 4/3$. The tank will be empty at 1:20 P.M.

34. Set up a coordinate system in which the tank has its lowest point at the origin and its vertical diameter lying on the y-axis, so that the equation of the cross section of the tank in the xy-plane will be $x^2 + (y - 4)^2 = 16$. If the liquid in the tank has depth y, then the radius at its surface is $x = \left(8y - y^2\right)^{1/2}$, so in Eq. (20) we take $A(y) = \pi x^2 \left(8y - y^2\right)$, $c = 1$, $g = 32$, and $a = \pi/144$ to obtain

$$A(y) = \pi x^2 = \pi\left(8y - y^2\right).$$

Thus

$$\pi\left(8y - y^2\right)\frac{dy}{dt} = -\frac{\pi}{144}\sqrt{64y} :$$
$$\left(8y^{1/2} - y^{3/2}\right) dy = -\frac{1}{18}\, dt;$$
$$\frac{16}{3}y^{3/2} - \frac{2}{5}y^{5/2} = -\frac{t}{18} + C.$$

When $t = 0$, $y = 8$. It follows that $C = \frac{512}{15}\sqrt{2}$. Thus

$$\frac{16}{3}y^{3/2} - \frac{2}{5}y^{5/2} = -\frac{t}{18} + \frac{512}{15}\sqrt{2}.$$

The tank is empty when $t = 0$. At that time we have $t = \frac{3072}{5}\sqrt{2} \approx 869$ seconds—about 14 minutes and 29 seconds.

Section 6.6

1. $W = \displaystyle\int_{-2}^{1} 10\, dx = 30.$

4. $W = \displaystyle\int_{0}^{4} -3\sqrt{x}\, dx = -16.$

7. $W = \displaystyle\int_{0}^{1} 30x\, dx = 15$ (ft-lb).

10. $W = \int_0^{10} (62.4)(y)(25\pi)\, dy = 78,000\pi \approx 245,044.23$ (ft-lb).

13. Take $y = -10$ as ground level, so that the radius x of a thin horizontal slab of water at height y above ground level satisfies the equation $x^2 = 5y$. Therefore

$$W = \int_0^5 (y+10)(50)(5\pi y)\, dy = \frac{125,000}{3}\pi \approx 130,900 \text{ (ft-lb)}.$$

16. Set up the following coordinate system: The x-axis and y-axis cross at the center of one end of the tank, so that the equation of the circular (vertical) cross section of the tank is $x^2 + y^2 = 9$. Then the gasoline must be lifted to the level $y = 10$. A horizontal cross section of the tank at the level y is a rectangle of length 10 and width $2x$ where x and y satisfy the equation $x^2 + y^2 = 9$ of the end of the tank. Thus we find that $x = (9 - y^2)^{1/2}$. So

(a) The amount of work required to pump all of the gasoline into the cars is:

$$W = \int_{-3}^3 \left(2\sqrt{9 - y^2}\right)(10)(10 - y)(45)\, dy$$

$$= 900 \int_{-3}^3 \left(10\sqrt{9 - y^2} - y\sqrt{9 - y^2}\right) dy$$

$$= 9000 \cdot \frac{1}{2} \cdot \pi \cdot 9 + 900 \left[\frac{1}{3}\left(9 - y^2\right)^{3/2}\right]_{-3}^3$$

$$= 40,500\pi \approx 127,234.5 \text{ (ft-lb)}.$$

(b) Assume that the tank has a 1.341 hp motor. If it were to operate at 100% efficiency, it would pump all of the gasoline in $40500\pi/((3300)(1.341)) \approx 2.875161$ min, using 1 kW for 2.875161 minutes. This would amount to 0.047919356 kWh, costing 0.345¢. Assuming that the pump in the tank is only 30% efficient, the actual cost would be about 1.15¢.

19. Take $y = 0$ at ground level. At time t, $0 \le t \le 50$, the bucket is at height $y = 2t$, and its weight is $F = 100 - \frac{1}{2}t$. So $t = \frac{1}{2}y$, and therefore the work is given by

$$W = \int_0^{100} \left(100 - \frac{1}{4}y\right) dy = 8750 \text{ (ft-lb)}.$$

22. $W = \int_{x_1}^{x_2} Ap(Ax)\, dx$. Let $V = Ax$; then $dV = A\, dx$. Therefore

$$W = \int_{V_1}^{V_2} pV\, dV.$$

25. $W = \int_0^1 60\pi(1 - y)\sqrt{y}\, dy = 16\pi$ (ft-lb).

28. There are a number of ways to work this problem. Here is a way to check your answer by using elementary physics and no calculus. Stage 1: Imagine the chain hanging in the shape of an "L" with 40 feet vertical and 10 feet horizontal, the latter on the floor of the monkey's cage in a neat heap. Stage 2: Cut off this ten-foot length of chain and move it to its final position— hanging from the top of the cage. It weighs 5 pounds and has moved an average distance of 45 feet, so the work to lift this segment of the chain is $S_1 = (5)(45) = 225$ ft-lb. Stage 3: Return to the dangling chain. Cut off its bottom 15 feet and move it to its final position—hanging from the 10-foot segment now hanging from the top of the cage. The top end of this segment is lifted $30 - 15 = 15$ feet and it weighs 7.5 pounds, so the work to lift the second segment is $S_2 = (15)(7.5) = 112.5$ ft-lb. The remaining 25 feet of the chain doesn't move at all. Finally, the monkey lifts her own 20 pounds a

distance of 40 feet, so the work involved here is $S_3 = 800$ ft-lb. So the total work in the process is $W = W_1 + W_2 + W_3 = 1137.5$ ft-lb. Of course, to gain the maximum benefit from this problem, you should work it using techniques of calculus.

31. Set up a coordinate system with the y-axis vertical and the x-axis coinciding with the bottom of one end of the trough. A horizontal section of the trough at y is $2 - y$ feet below the water surface, so the total force on the end of the trough is given by

$$F = \int_0^2 (2)(2-y)(\rho)\, dy = 6\rho = 249.6 \text{ (pounds)}.$$

34. Describe the end of the tank by the inequality $x^2 + y^2 \leq 16$, so that a horizontal section at level y has width $2x = 2\left(16 - y^2\right)^{1/2}$. Then the total force on the end of the tank is

$$F = \int_{-4}^4 \rho(2)(4-y)\left(16 - y^2\right)^{1/2} dy$$

$$= 2\rho \int_{-4}^4 4\left(16 - y^2\right)^{1/2} dy - 2\rho \int_{-4}^4 y\left(16 - y^2\right)^{1/2} dy$$

$$= (16\rho)(4\pi) = 64\rho\pi = 3200\pi \approx 10053.1 \text{ (pounds)}.$$

37. Place the x-axis along the water surface. Then a horizontal section of the triangle at level y has width $2x = \frac{8}{5}(y + 15)$, and hence the total force on the gate is

$$F = \int_{-15}^{-10} -\frac{8}{5}\rho y(y + 15)\, dy.$$

Chapter 6 Miscellaneous

1. Note that $t^2 - t - 2 = (t+1)(t-2)$ is negative on the interval $(-1, 2)$. Therefore the net distance traveled is

$$\int_0^3 (t^2 - t - 2)\, dt = -\tfrac{3}{2},$$

while the total distance traveled is

$$-\int_0^2 (t^2 - t - 2)\, dt + \int_2^3 (t^2 - t - 2)\, dt = \tfrac{31}{6}.$$

4. $V = \int_0^1 x^3\, dx = \tfrac{1}{4}.$

7. $V = \int_0^1 \pi\left(x^2 - x^4\right) dx = \pi\left(\tfrac{1}{3} - \tfrac{1}{5}\right) = \tfrac{2}{15}\pi.$

10. $V = \int_0^1 \left(2x - x^2 - x^3\right)^2 dx = \tfrac{22}{105}.$

13. $M = \int_0^{20} \frac{\pi}{16}(8.5)\, dx = (10.625)\pi \approx 33.379$ (grams).

16. Because $(a - h, r)$ lies on the ellipse, $\left(\dfrac{a-h}{a}\right)^2 + \left(\dfrac{r}{b}\right)^2 = 1$. Therefore $r^2 = \dfrac{2ah - h^2}{a^2} b^2$. And so

$$V = \int_{a-h}^a \pi y^2\, dx = \int_{a-h}^a \pi b^2\left(1 - \frac{x^2}{a^2}\right) dx = \pi\frac{b^2 h^2}{3a^2}\left(3a - h\right).$$

But $r^2 = \dfrac{b^2}{a^2} h(2a - h)$, so $\dfrac{b^2}{a^2} h = \dfrac{r^2}{2a - h}$. Therefore

$$V = \tfrac{1}{3}\pi r^2 h \, \frac{3a - h}{2a - h}.$$

19. $V = \displaystyle\int_1^t \pi \left(f(x)\right)^2 \, dx = \frac{\pi}{6}\left((1 + 3t)^2 - 16\right)$. Thus

$$\pi \left(f(x)\right)^2 = \frac{\pi}{6}\left((2)(1 + 3x)(3)\right) = \pi(1 + 3x).$$

Therefore $f(x) = \sqrt{1 + 3x}$.

22. If $-1 \le x \le 2$, then a thin vertical strip of the region above x is rotated in a circle of radius $x + 2$. Therefore the volume generated is

$$V = \int_{-1}^{2} 2\pi(x + 2)(x + 2 - x^2) \, dx = \tfrac{45}{2}\pi.$$

25. $\dfrac{dx}{dy} = \dfrac{3}{8}\left(\dfrac{4}{3}y^{1/3} - \dfrac{4}{3}y^{-1/3}\right)$. So—after some algebra—

$$\left(1 + \left(\frac{dx}{dy}\right)^2\right)^{1/2} = \tfrac{1}{2}\left(y^{1/3} + y^{-1/3}\right).$$

Therefore

$$L = \tfrac{1}{2}\int_1^8 \left(y^{1/3} + y^{-1/3}\right) \, dy = \tfrac{63}{8}.$$

28. $\dfrac{dy}{dx} = -\dfrac{x}{\sqrt{r^2 - x^2}}$, so $1 + \left(\dfrac{dy}{dx}\right)^2 = \dfrac{r^2}{r^2 - x^2}$. Therefore

$$A = \int_a^b 2\pi \sqrt{r^2 - x^2} \, \frac{r}{\sqrt{(r^2 - x^2}} \, dx = 2\pi r h.$$

34. $(y + 1)^{-1/2} \, dy = 1 \, dx$;

$2(y + 1)^{1/2} = x + C$;

$y(x) = -1 + \tfrac{1}{4}(x + C)^2$.

37. $y^2 \, dy = \dfrac{1}{x^2} \, dx$;

$\tfrac{1}{3}y^3 = -\dfrac{1}{x} + K$;

$y^3 = C - \tfrac{3}{4}$;

$y(x) = \left(C - \dfrac{3}{x}\right)^{1/3}.$ $\qquad 1 = y(1) = (C - 3)^{1/3}$: $\quad C = 4$.

$y(x) = \left(4 - \dfrac{3}{x}\right)^{1/3}.$

40. $y^{-1/2} \, dy = \sin x \, dx$;

$2y^{1/2} = C - \cos x$;

$y(x) = \tfrac{1}{4}(C - \cos x)^2.$ $\qquad 4 = y(0) = \tfrac{1}{4}(C - 1)^2$: $\quad C = 5$ or $C = -3$.

Two solutions: $y(x) = \tfrac{1}{4}(5 - \cos x)^2$, $\; y(x) = \tfrac{1}{4}(3 + \cos x)^2$.

Chapter 6 Miscellaneous

43. Denote by K the spring constant. Then

$$\int_2^5 K(x - L)\,dx = 5\int_2^3 K(x - L)\,dx.$$

After applying the fundamental theorem of calculus, we find that

$$(5 - L)^2 - (2 - L)^2 = 5(3 - L)^2 - 5(2 - L)^2.$$

Solve for L: $L = 1$, and therefore the natural length of the spring is 1 foot.

46. Set up a coordinate system with the axis of the cone lying on the y-axis and with a diameter of the cone lying on the x-axis. Now a horizontal slice of the cone at height y has radius given by $x = \frac{1}{2}(1 - y)$; the units here are in feet. Therefore the work done in building the anthill is

$$W = \int_0^1 \tfrac{1}{4}(150y)\pi(1 - y)^2\,dy = \tfrac{25}{8}\pi \approx 9.82 \quad \text{(ft-lb)}.$$

49. If the coordinate system is chosen with the origin at the midpoint of the bottom of the dam and with the x-axis horizontal, then the equation of the slanted edge of the dam is $y = 2x - 200$ (with units in feet). Therefore the width of the dam at level y is $2x = y + 200$. Thus the total force on the dam is

$$F = \int_0^{100} \rho(100 - y)(y + 200)\,dy.$$

Chapter 7: Exponential and Logarithmic Functions

Section 7.1

1. $2^3 \cdot 2^4 = 2^{3+4} = 2^7 = 128.$

4. $2^{2^3} = 2^8 = 256.$

7. $\left(2^{12}\right)^{1/3} = 2^{(12/3)} = 2^4 = 16.$

10. $6^5 \cdot 3^{-5} = 2^5 \cdot 3^5 \cdot 3^{-5} = 2^5 \cdot 3^0 = 32.$

13. $\log_5 125 = \log_5 5^3 = 3\log_5 5 = 3.$

16. $\log_{12} 12^2 = 2\log_{12} 12 = 2.$

19. $\ln 6 = \ln(2 \cdot 3) = (\ln 2) + (\ln 3).$

22. $\ln 200 = \ln(2^3 \cdot 5^2) = (3\ln 2) + (2\ln 5).$

25. $\ln \frac{27}{40} = \ln(3^3) - \ln(2^3 \cdot 5).$

28. If $x = \log_{0.5} 16$, this means that $(0.5)^x = 16$. Take the reciprocal of each side in the last equation to obtain
$$2^x = \frac{1}{16} = \frac{1}{2^4} = 2^{-4},$$
then take the base 2 logarithm of the first and last terms above to conclude that $x = -4$. Finally verify this answer by substitution (because the method is predicated on the assumption that x exists—that is, that the logarithm to the base 1/2 of 16 exists).

34. $3^{2x} = 3^4$: $2x = 4$, so $x = 2$.

40. $2e^{-7x} = 5$: $e^{-7x} = \frac{5}{2}$; $-7x = \ln 5 - \ln 2$; $x = -\frac{1}{7}(\ln 5 - \ln 2).$

43. $\dfrac{dy}{dx} = x^{1/2}e^x + \frac{1}{2}x^{-1/2}e^x$

46. $\dfrac{dy}{dx} = \dfrac{2x - 1}{2x^{3/2}}e^x$

49. $\dfrac{dy}{dx} = \frac{1}{2}x^{1/2}\ln x + x^{-1/2}$

52. $\dfrac{dy}{dx} = e^x\left(\dfrac{1}{x} + \ln x\right)$

58. $f'(x) = -10e^{-10x}.$

61. If $P(t) = 3^t$ then $P(t) = \left(e^{\ln 3}\right)^t = e^{t\ln 3}$, so $P'(t) = (\ln 3)\,e^{t\ln 3} = (\ln 3)\,(3^t) = 3^t\ln 3.$

64. $P'(t) = 2^t \ln 2.$

    (a) $P'(t) = \ln 2$, about 0.69315 (millions/hour).

    (b) $P'(4) = 2^4 \ln 2 = 16\ln 2$, approximately 11.09035 (millions/hour).

Section 7.2

1. $f'(x) = \dfrac{1}{3x - 1}D_x(3x - 1) = \dfrac{3}{3x - 1}$

4. $f(x) = 3\ln(1 + x)$, so $f'(x) = \dfrac{3}{1 + x}.$

7. $f'(x) = -\dfrac{1}{x}\sin(\ln x)$

10. $f'(x) = \dfrac{1}{\ln x}D_x(\ln x) = \dfrac{1}{x\ln x}$

16. $f'(x) = (\cos(\ln 2x))\,D_x(\ln 2x) = \dfrac{1}{x}\cos(\ln 2x)$

19. $f(x) = 3\ln(2x+1) + 4\ln(x^2-4)$, so $f'(x) = \dfrac{6}{2x+1} + \dfrac{8x}{x^2-4}$.

22. $f(x) = \frac{1}{2}\ln(4x-7) - 3\ln(3x-2)$, so $f'(x) = \dfrac{2}{4x-7} - \dfrac{9}{3x-2} = \dfrac{59-30x}{(4x-7)(3x-2)}$.

25. *Suggestion:* Write $g(t) = 2\ln t - \ln(t^2+1)$.

28. $f(x) = \ln(\sin x) - \ln(\cos x)$, so $f'(x) = \dfrac{\cos x}{\sin x} + \dfrac{\sin x}{\cos x} = \cot x + \tan x$.

34. $\displaystyle\int \dfrac{dx}{3x+5} = \frac{1}{3}\ln|3x+5| + C$.

40. $\displaystyle\int \dfrac{1}{x\ln x}\,dx = \ln|\ln x| + C$.

46. $\displaystyle\int \dfrac{\ln(x^3)}{x}\,dx = 3\int \dfrac{\ln x}{x}\,dx = \frac{3}{2}(\ln x)^2 + C$.

49. The numerator is $\frac{1}{3}$ the derivative of the denominator.

52. $\displaystyle\lim_{x\to\infty} \dfrac{\ln x^3}{x^2} = \lim_{x\to\infty} \dfrac{3\ln x}{x^2} = 0$.

55. *Suggestion:* Make the substitution $x = \dfrac{1}{u^2}$.

58. $f^{(1)}(x) = x^{-1}$, $f^{(2)}(x) = (-1)x^{-2}$, $f^{(3)}(x) = (-2)(-1)x^{-3}$, $f^{(4)}(x) = (-3)(-2)(-1)x^{-4}$, ..., and, in general, $f^{(n)}(x) = (-1)^{n-1}(n-1)!\,x^{-n}$ for $n \geq 1$. Proof: By induction on n.

64. Given: $y = f(x) = x\ln x$ for $x > 0$. Then $f'(x) = 1 + \ln x$ and $f''(x) = 1/x$. The critical point occurs where $\ln x = -1$, so $x = 1/e$ and $y = (1/e)\ln(1/e) = -1/e$. There are no possible points of inflection. The graph is decreasing on $(0, 1/e)$ and is increasing if $x > 1/e$. It is concave upward everywhere, and its only intercept is $(1,0)$. As $x \to 0^+$, $y \to 0$; moreover, $dy/dx \to -\infty$. This helps sketch the graph near the point $(0,0)$.

67. After simplifications, you will find that $\dfrac{dy}{dx} = \dfrac{2-\ln x}{2x^{3/2}}$ and $\dfrac{d^2y}{dx^2} = \dfrac{3\ln x - 8}{4x^{5/2}}$.

70. $\dfrac{dy}{dx} = -\dfrac{1}{x^2}$; $\left(1 + \left(\dfrac{dy}{dx}\right)^2\right)^{1/2} = (1+x^{-4})^{1/2} = \dfrac{(x^4+1)^{1/2}}{x^2}$. The surface area of the part of the horn over the interval from $x = 1$ to $x = b$ (for $b > 1$) is then

$$A_b = \int_1^b \dfrac{2\pi}{x^3}(x^4+1)^{1/2}\,dx \geq \int_1^b \dfrac{2\pi}{x^3}x^2\,dx = 2\pi\ln b.$$

Therefore the surface area of Gabriel's horn is infinite because $A_b \to +\infty$ as $b \to +\infty$. Next,

$$V_b = \int_1^b \dfrac{\pi}{x^2}\,dx = \pi\left(1 - \dfrac{1}{b}\right) \to \pi \text{ as } b \to +\infty.$$

In summary, the volume is finite although the surface area is infinite.

Section 7.3

1. $f'(x) = e^{2x}D_x(2x) = 2e^{2x}$

4. $f'(x) = -3x^2 e^{4-x^3}$

10. $g'(t) = \frac{1}{2}(e^t - e^{-t})^{-1/2}(e^t + e^{-t})$

16. $f'(x) = e^x\cos 2x - 2e^x\sin 2x$

22. $g'(t) = \dfrac{2e^t}{(1-e^t)^2}$

28. $f'(x) = (e^{2x} + e^{-2x})^{-1/2}(e^{2x} - e^{-2x})$

34. $\dfrac{dy}{dx} = \dfrac{1}{(1+y)\,e^y}$

40. $\displaystyle\int \sqrt{x}\,e^{2x\sqrt{x}} = \tfrac{1}{3}e^{2x\sqrt{x}} + C$

46. $\displaystyle\int e^{2x+3}\,dx = \tfrac{1}{2}e^{2x+3} + C$

49. Suggestion: Let $u = \sqrt{x}$.

52. Because $e^{a+b} = e^a e^b$, we have $\displaystyle\int e^{x+e^x}\,dx = \int e^x\,e^{e^x}\,dx = e^{e^x} + C$. Note that e^{e^x} means $e^{(e^x)}$.

58. $\displaystyle\lim_{h\to 0} (1+2h)^{1/h} = e^2$

64. First, $dy/dx = 3x^2 e^{-x} - x^3 e^{-x} = x^2(3-x)\,e^{-x}$
and $d^2y/dx^2 = x(x^2 - 6x + 6)\,e^{-x}$. Note that $y = 0$
when $x = 0$ and that this is the only intercept. Next,
$dy/dx = 0$ when $x = 0$ and when $x = 3$, so there are
horizontal tangents at $(0,0)$ and at $\left(3, \dfrac{27}{e^3}\right)$—near

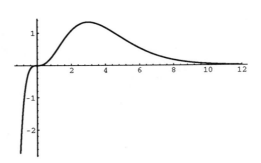

the point $(3, 1.344)$. The second derivative vanishes
when $x = 0$ and when $x^2 - 6x + 6 = 0$; the latter
equation has the two solutions $x = 3 \pm \sqrt{3}$. Consequently there may be inflection points at $(0,0)$, near
$(1.268, 0.574)$, and near $(4.732, 0.933)$. The graph is
increasing if $x < 3$ and decreasing if $x > 3$. It is concave downward if $x < 0$ and if $|x - 3| < \sqrt{3}$,
concave upward if **both** $|x - 3| > \sqrt{3}$ and $x > 0$. Because $x^3 e^{-x} \to 0$ as $x \to +\infty$, the x-axis is a
horizontal asymptote. The graph is shown above.

67. $V = \displaystyle\int_0^1 \pi e^{2x}\,dx = \dfrac{\pi}{2}\left(e^2 - 1\right) \approx 10.0359.$

70. $A = \displaystyle\int_0^1 (2\pi)\left(\tfrac{1}{2}\left(e^x + e^{-x}\right)\right)\left(\tfrac{1}{2}\left(e^x + e^{-x}\right)\right)\,dx = \dfrac{\pi}{2}\int_0^1 \left(e^{2x} + 2 + e^{-2x}\right)\,dx$

$= \dfrac{\pi}{4}\left(e^2 + 4 - e^{-2}\right) \approx 8.83865.$

73. $f'(x) = (n-x)x^{n-1}e^{-x}$ and $f''(x) = \left((x-n)^2 - n\right)x^{n-2}e^{-x}$. Because $f(x) \geq 0$ for $x \geq 0$, $f(0) = 0$,
and $f(x) \to 0$ as $x \to +\infty$, $f(x)$ must have a maximum, and the critical point where $x = n$ is the
sole candidate. Evaluate $f(n)$ to obtain the global maximum value of $f(x)$.

76. To avoid subscripts, write p for m_1, q for m_2, A for C_1, and B for C_2. Then with

$$y(x) = Ae^{px} + Be^{qx}, \text{ we obtain}$$
$$y'(x) = Ape^{px} + Bqe^{qx} \text{ and}$$
$$y''(x) = Ap^2 e^{px} + Bq^2 e^{qx}.$$

So $ay''(x) + by'(x) + cy(x) = \left(aAp^2 + bAp + cA\right)e^{px} + \left(aBq^2 + bBq + cB\right)e^{qx}$
$$= A\left(ap^2 + bp + c\right)e^{px} + B\left(aq^2 + bq + c\right)e^{qx}$$
$$= (A)(0)e^{px} + (B)(0)e^{qx} = 0.$$

Section 7.4

1. $f'(x) = 10^x \ln 10$

4. $f(x) = (\log_{10} e)(\ln \cos x) = \dfrac{1}{\ln 10}\ln(\cos x)$, so $f'(x) = \left(\dfrac{1}{\ln 10}\right)\left(\dfrac{-\sin x}{\cos x}\right)$.

10. $f'(x) = 7^{(8^x)}(\ln 7)(8^x \ln 8)$

16. $f(x) = (\log_2 e)(\log_e x) = \dfrac{\ln x}{\ln 2}$, so $f'(x) = \dfrac{1}{x \ln 2}$.

22. $f'(x) = \pi^x \ln \pi + \pi x^{\pi - 1}$

28. The substitution $u = 1/x$ yields

$$\int \frac{10^{1/x}}{x^2}\, dx = \int -10^u\, du = -\frac{10^u}{\ln 10} + C = -\frac{10^{1/x}}{\ln 10} + C.$$

31. Write $\log_2 x$ as $(\ln x)/(\ln 2)$. Then, if you wish, use the substitution $u = \ln x$.

34. $\ln y = \frac{1}{3}\ln(3 - x^2) - \frac{1}{4}\ln(x^4 + 1)$, so

$$\frac{1}{y} \cdot \frac{dy}{dx} = \frac{-2x}{3(3 - x^2)} - \frac{x^3}{x^4 + 1};$$

$$\frac{dy}{dx} = -\frac{(3 - x^2)^{1/2}}{(x^4 + 1)^{1/4}}\left(\frac{2x}{3(3 - x^2)} + \frac{x^3}{x^4 + 1}\right).$$

37. $\ln y = (\ln x)^2$. So $\dfrac{dy}{dx} = (y)\,\dfrac{2\ln x}{x} = \left(x^{\ln x}\right)\dfrac{2\ln x}{x}$.

40. $\dfrac{dy}{dx} = (y)\left(\dfrac{1}{2(x + 1)} + \dfrac{1}{3(x + 2)} + \dfrac{1}{4(x + 3)}\right)$

$$= \sqrt{x + 1}\sqrt[3]{x + 2}\sqrt[4]{x + 3}\left(\frac{1}{2(x + 1)} + \frac{1}{3(x + 2)} + \frac{1}{4(x + 3)}\right).$$

46. $\ln y = x \ln\left(1 + \dfrac{1}{x}\right) = x \ln(x + 1) - x \ln x$.

$$\frac{dy}{dx} = \left(1 + \frac{1}{x}\right)^x \left(\ln(x + 1) + \frac{x}{x + 1} - \ln x - 1\right).$$

52. $\ln y = x \ln(\cos x)$. Therefore $\dfrac{dy}{dx} = (\cos x)^x\left(\ln(\cos x) - \dfrac{x \sin x}{\cos x}\right)$.

55. If $y = \ln\left(a^{1/x}\right)$ then $y = \dfrac{1}{x}\ln a$.

$$\lim_{x \to \infty} y = \lim_{x \to \infty} \frac{1}{x}(\ln a) = 0.$$

$$\lim_{x \to \infty} \ln\left(a^{1/x}\right) = 0,$$

$$\lim_{x \to \infty} a^{1/x} = e^0 = 1.$$

58. $\displaystyle\lim_{x \to 0^+} \frac{1}{1 + 2^{1/x}} = \lim_{k \to \infty} \frac{1}{1 + 2^k} = 0.$ $\displaystyle\lim_{x \to 0^-} \frac{1}{1 + 2^{1/x}} = \lim_{k \to -\infty} \frac{1}{1 + 2^k} = 1.$

Section 7.5

1. The principal at time t is $P(t) = 1000e^{(0.08)t}$. So $P'(t) = 80e^{(0.08)t}$.

   Answers: $P'(5) = \$119.35$ and $P'(20) = \$396.24$.

4. Suppose that the skull was formed at time $t = 0$ (in years). Then the amount of ^{14}C it contains at time t will be

$$Q(t) = Q_0 e^{-kt}$$

where Q_0 is the initial amount and $k = \dfrac{\ln 2}{5700} \approx 0.0001216$. We find the value $t = T$ corresponding to "now" by solving

$$Q(t) = \tfrac{1}{6}Q_0 : \quad \tfrac{1}{6}Q_0 = Q_0 e^{-kT}, \text{ so } 6 = e^{kT};$$

therefore

$$T = \frac{1}{k}\ln 6 = 5700\,\frac{\ln 6}{\ln 2} \approx 14{,}734,$$

the approximate age of the skull in years.

7. $1 + r = \left(1 + \dfrac{0.09}{4}\right)^4$ for quarterly compounding, and in this case we find that (a) $r \approx 9.308\%$. With the aid of similar formulas, we obtain the other four answers.

10. If $C(t)$ is the concentration of the drug at time t (hours), then

$$C(t) = C_0 e^{-kt}$$

where $k = \tfrac{1}{5}\ln 2$. We require C_0 so large that $C(1) = (45)(50) = 2250$. Thus $C_0 e^{-k} \geq 2250$; that is, $C_0 \geq 2250 e^k \approx 2584.57$ (mg).

13. Let Q denote the amount of radioactive cobalt remaining at time t (in years), with the occurrence of the accident set at time $t = 0$. Then

$$Q = Q_0 e^{-(t\ln 2)/(5.27)}.$$

If T is the number of years until the level of radioactivity has dropped to a hundredth of its initial value, then

$$\frac{1}{100} = e^{-(T\ln 2)/(5.27)},$$

and it follows that $T = (5.27)\,\dfrac{\ln 100}{\ln 2} \approx 35$ (years).

16. Let $T = T(t)$ denote the temperature at time t; by Newton's law of cooling, we have

$$\frac{dT}{dt} = k(T - A).$$

Here, $A = 0$, $T(0) = 25$, and $T(20) = 15$. Also $\dfrac{dT}{dt} = kT$, so

$$T(t) = T_0 e^{kt} = 25 e^{kt}.$$

Next, $15 = T(20) = 25 e^{20k}$, so $k = \tfrac{1}{20}\ln\left(\tfrac{3}{5}\right)$. Now $T(t) = 5$ when $5 = 25 e^{kt}$; that is, when

$$t = -\frac{20\ln 5}{\ln\left(\tfrac{3}{5}\right)} \approx 63.01.$$

Answer: The buttermilk will be at $5°C$ about one hour and three minutes after putting it on the porch.

19. We begin with the equation $p(x) = (29.92)e^{-x/5}$.

(a) $p\left(\dfrac{10{,}000}{5280}\right) \approx 20.486$ (inches); $p\left(\dfrac{30{,}000}{5280}\right) \approx 9.604$ (inches).

(b) If x is the altitude in question, then we must solve

$$15 = (29.92)e^{-x/5};$$
$$x = 5\ln\left(\frac{29.92}{15}\right) \approx 3.4524 \text{ (miles)},$$

approximately $18{,}230$ feet.

1. Given: $\dfrac{dy}{dx} = y + 1$; $y(0) = 1$.

$$\frac{1}{y+2}\frac{dy}{dx} = 1;$$
$$\ln(y+1) = x + C;$$
$$y + 1 = Ke^x.$$

When $x = 0$, $y = 1$. Therefore $2 = (K)(1) = K$, so the solution is $y(x) = -1 + 2e^x$.

4. $\dfrac{dy}{dx} = \dfrac{1}{4} - \dfrac{y}{16} = -\dfrac{1}{16}(y - 4)$; $y(0) = 20$.

$$\frac{1}{y-4}\frac{dy}{dx} = -\frac{1}{16};$$
$$\ln(y-4) = C_1 - \frac{x}{16};$$
$$y = 4 + Ce^{-x/16}.$$

We are given $20 = y(0) = 4 + C$, so $C = 16$. Answer: $y(x) = 4 + 16e^{-x/16}$.

10. $v(t) = 10 - 20e^{5t}$

16. Let $Q(t)$ be the number of pounds of salt in the tank at time t (in seconds). Now $Q(0) = 50$, and

$$\frac{dQ}{dt} = -\frac{5}{1000}Q(t) = -\frac{1}{200}Q(t).$$

Therefore

$$Q(t) = 50e^{-t/200}.$$

Next, $Q(t) = 10$ when $e^{-t/200} = \frac{1}{5}$, so that $t = 200\ln 5$. So 10 pounds of salt will remain in the tank after approximately 5 minutes and 22 seconds.

19. We have $v = v_0 e^{-kt}$ where we are given $v_0 = 40$. Here, also, $v(10) = 20$, so that $20 = 40e^{-10k}$. It follows that $k = (0.1)\ln 2$, so the total distance traveled will be $\dfrac{1}{k}v_0 = \dfrac{(40)(10)}{\ln 2}$, approximately 577 feet.

22. $\dfrac{dx}{dt} + ax = be^{ct}$, $x(0) = x_0$, $a + c \neq 0$.

$$e^{at}\frac{dx}{dt} + axe^{at} = be^{(a+c)t}.$$
$$D_t\left(e^{at}x(t)\right) = be^{(a+c)t}.$$
$$e^{at}x(t) = C + \frac{b}{a+c}e^{(a+c)t} \qquad (a + c \neq 0).$$

So $x(t) = Ce^{-at} + \dfrac{b}{a+c}e^{ct}$, and $x_0 = x(0) = C + \dfrac{b}{a+c}$, and hence $C = x_0 - \dfrac{b}{a+c}$.

Therefore $x(t) = \left(x_0 - \dfrac{b}{a+c}\right)e^{-at} + \dfrac{b}{a+c}e^{ct} = x_0 e^{-at} + \dfrac{b}{a+c}(e^{ct} - e^{-at})$.

25. $\dfrac{dx}{dt} = k(100000 - x(t))$:

$$\frac{-dx}{100000 - x} = -k\,dt;$$
$$\ln(100000 - x) = C_1 - kt;$$
$$100000 - x = Ce^{-kt};$$
$$x(t) = 100000 - Ce^{-kt}.$$

On March 1, $t = 0$ and $x = 20000$. On March 15, $t = 14$ and $x = 60000$.

$$20000 = 100000 - C, \text{ so } C = 80000$$
$$x(t) = 10000\left(10 - 8e^{-kt}\right).$$
$$60000 = x(14) = 10000(10 - 8e^{-14k}) \text{ so } 6 = 10 - 8e^{-14k}. \text{ Solve for } k: k = \tfrac{1}{14}\ln 2.$$

(a) $x(t) = 10000(10 - 8e^{-kt})$ where $k = \tfrac{1}{14}\ln 2$.

(b) $x(T) = 80000$: Solve $10 - 8e^{-kT} = 2$ for T: $T = \dfrac{1}{k}\ln 4 = 28$. So $80,000$ people will be infected on March 29.

(c) $\lim\limits_{t\to\infty} N(t) = 100,000$: Eventually everybody gets the flu.

Chapter 7 Miscellaneous

1. $f(x) = \ln 2\sqrt{x} = \ln 2 + \tfrac{1}{2}\ln x$, so $f'(x) = \dfrac{1}{2x}$.

4. $f'(x) = \tfrac{1}{2}\left(x^{-1/2}\right) 10^{\left(x^{1/2}\right)}\ln 10$

7. $f'(x) = 3x^2 e^{-1/x^2} + x^3 e^{-1/x^2}\left(2x^{-3}\right) = \left(2 + 3x^2\right)e^{-1/x^2}$.

10. $f'(x) = (\exp(10^x))(10^x)(\ln 10)$

13. $f'(x) = \dfrac{(x-1)-(x+1)}{(x-1)^2}\exp\left(\dfrac{x+1}{x-1}\right) = -\dfrac{2}{(x-1)^2}\exp\left(\dfrac{x+1}{x-1}\right)$

16. $f'(x) = \dfrac{1}{x}\cos(\ln x)$

19. $f'(x) = \dfrac{3^x\cos x + (3^x\ln 3)\sin x}{3^x\sin x} = \dfrac{\cos x + (\ln 3)\sin x}{\sin x} = \cot x + \ln 3$.

22. If $y = x^{\sin x}$ then $\ln y = (\sin x)(\ln x)$.

    So $\dfrac{dy}{dx} = y\left(\dfrac{1}{x}\sin x + (\cos x)(\ln x)\right) = \left(x^{\sin x}\right)\left(\dfrac{1}{x}\sin x + (\cos x)(\ln x)\right)$.

28. $\displaystyle\int \dfrac{e^x - e^{-x}}{e^x + e^{-x}}\, dx = \ln\left(e^x + e^{-x}\right) + C$.

31. Let $u = \sqrt{x}$.

34. Let $u = \ln x$. Then $du = \dfrac{1}{x}\, dx$, and so

    $$\int \dfrac{1}{x}(1 + \ln x)^{1/2}\, dx = \int (1 + u)^{1/2}\, du = \dfrac{2}{3}(1 + u)^{3/2} + C = \dfrac{2}{3}(1 + \ln x)^{3/2} + C.$$

37. First, $dx = 2t\, dt$, and so $x = t^2 + C$. But $x(0) = 17$, and so $x(t) = t^2 + 17$.

40. $e^{-x}\, dx = dt$, so $-e^{-x} = C + t$. Now $-e^{-2} = C + 0$, and so

    $$-e^{-x} = -e^{-2} + t;$$
    $$e^{-x} = e^{-2} - t;$$
    $$-x = \ln\left(e^{-2} - t\right);$$
    $$x(t) = -\ln\left(e^{-2} - t\right).$$

43. $\dfrac{dx}{x} = \cos t\, dt$, so

    $$\ln|x| = C_1 + \sin t.$$

    Consequently $x(t) = Ce^{\sin t}$. Because $x(0) = \sqrt{2}$, $C = \sqrt{2}$. So $x(t) = \sqrt{2}e^{\sin t}$.

46. $\dfrac{dy}{dx} = 1 - \dfrac{1}{x}$ and $\dfrac{d^2y}{dx^2} = \dfrac{1}{x^2}$. The graph is concave upward for all x, and there is a horizontal tangent at $(1,1)$. The function is decreasing on $(0,1)$ and increasing for $x > 1$. In addition, as $x \to 0^+$, $y(x) \to +\infty$ while $dy/dx \to -\infty$. The graph is shown on the right.

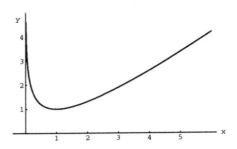

49. $\dfrac{dy}{dx} = \dfrac{1}{x^2}e^{-1/x}$ and $\dfrac{d^2y}{dx^2} = \dfrac{1 - 2x}{x^4}e^{-1/x}$. The graph is shown in the answer section of the text.

52. At time t (in years), the amount to be repaid will be $1000e^{(0.1)t}$. So the profit on selling would be

$$P(t) = 800e^{\left(\frac{1}{2}\sqrt{t}\right)} - 1000e^{(t/10)}.$$
$$P'(t) = 200t^{-1/2}e^{\left(\frac{1}{2}\sqrt{t}\right)} - 100e^{(t/10)}.$$

$P'(t) = 0$ when $2e^{\left(\frac{1}{2}\sqrt{t}\right)} = t^{1/2}e^{(t/10)}$. The iteration

$$t \longleftarrow \left(2e^{\left(\frac{1}{2}\sqrt{t}\right)}e^{(-t/10)}\right)^2$$

yields $t \approx 11.7519$ years as the optimal time to cut and sell. The resulting profit would be $1202.37.

58. Let $x(t)$ represent the position of the race car at time t, and let $x(0) = 0$. Then the velocity of the car is $v(t) = x'(t)$. We want to find $v(0) = v_0$.

$$\frac{dv}{dt} = -kv,$$
$$v(t) = v_0e^{-kt}, \text{ as usual.}$$
$$x(t) = C - \frac{v_0}{k}\left(e^{-kt}\right).$$

Since $0 = x(0) = C - \dfrac{v_0}{k}$, $C = \dfrac{v_0}{k}$, hence $x(t) = \dfrac{v_0}{k}\left(1 - e^{-kt}\right)$.

When $t = 0$, $\dfrac{dv}{dt} = -2$, so $-kv_0 = -2$, hence $k = \dfrac{2}{v_0}$ and so $x(t) = \dfrac{(v_0)^2}{2}\left(1 - e^{-2t/v_0}\right)$.

$$\lim_{t\to\infty} v(t) = 0, \text{ so } \lim_{t\to\infty} x(t) = 1800.$$

$$1800 = \lim_{t\to\infty}\frac{(v_0)^2}{2}\left(1 - e^{-2t/v_0}\right) = \frac{(v_0)^2}{2}: \quad (v_0)^2 = 3600; \quad v_0 = 60.$$

The initial velocity of the car was 60 m/s.

61. Let $Q(t)$ denote the temperature at time t, with t in hours and $t = 0$ corresponding to 11:00 P.M., the time of the power failure. Let $A = 20$ be the room temperature. By Newton's law of cooling,

$$\frac{dQ}{dt} = k(A - Q), \text{ so that } \frac{1}{A - Q}dQ = k\,dt$$

where k is a positive constant. Next, $\ln(A - Q) = C_1 - kt$, so that $Q(t) = A - Ce^{-kt}$. Now $Q_0 = Q(0) = A - C$, so $C = A - Q_0$. Thus

$$Q(t) = A + (Q_0 - A)e^{-kt}.$$

We are given $Q_0 = -16$, $A = 20$, and $Q(7) = -10$. We must find the value $t = T$ at which $Q(T) = 0$. Now

$$Q(t) = 20 - 36e^{-kt};$$
$$Q(7) = -10 = 20 - 36e^{-7k};$$
$$30 = 36e^{-7k};$$
$$k = \tfrac{1}{7}\ln(1.2).$$

Next, $0 = Q(T) = 20 - 36e^{-kT}$; $20 = 36e^{-kT}$: Therefore $e^{kT} = 1.8$, and thus

$$T = \frac{1}{k}\ln(1.8) = \frac{7\ln(1.8)}{\ln(1.2)} \approx 22.5673.$$

The critical temperature will be reached about 22 hours and 34 minutes after the power goes off; that is, at about 9:34 P.M. on the following day. (The data used in this problem are those obtained during an actual incident of the sort described.)

64. Let S be the safe limit, and let $R(t)$ be the radiation level at time t (in years).

$$R(t) = 10Se^{-kt}.$$

$R(\tfrac{1}{2}) = 95 = 105e^{-k/2}$; $\tfrac{9}{10} = e^{-k/2}$; $\dfrac{k}{2} = \ln\tfrac{10}{9}$; $k = 2\ln\tfrac{10}{9}$.

$R(t) = S$ when $10Se^{-kt} = S$: $e^{kt} = 10$; $t = \dfrac{\ln 10}{k} = \dfrac{\ln 10}{2\ln\left(\tfrac{10}{9}\right)} \approx 10.927$. Radiation will drop to a safe level in just under 11 years.

4. (a) $\sec^{-1}(1) = 0$ (b) $\sec^{-1}(-1) = \pi$ (c) $\sec^{-1}(2) = \pi/3$ (d) $\sec^{-1}(-\sqrt{2}) = 3\pi/4$

10. $f'(x) = \dfrac{x}{1+x^2} + \arctan x$

16. $f'(x) = \dfrac{1}{2x\sqrt{x-1}}$

22. $f'(x) = \dfrac{e^{\arcsin x}}{(1-x^2)^{1/2}}$

28. $\dfrac{dy}{dx} = -\left(1-y^2\right)^{1/2}\left(1-x^2\right)^{-1/2}$; Tangent: $x\sqrt{3} + 3y = 2\sqrt{3}$

34. $\displaystyle\int_{-2}^{-2/\sqrt{3}} \dfrac{dx}{x\sqrt{x^2-1}} = \operatorname{arcsec}\left|\dfrac{-2}{\sqrt{3}}\right| - \operatorname{arcsec}|-2| = \dfrac{\pi}{6} - \dfrac{\pi}{3} = -\dfrac{\pi}{6}$

37. Suggestion: Let $u = 2x$.

40. Let $u = \frac{2}{3}x$, so that $2x = 3u$ and $dx = \frac{3}{2}\,du$. Then
$$\int \frac{dx}{x\left(4x^2-9\right)^{1/2}} = \int \frac{(3/2)\,du}{(3/2)u\left(9u^2-9\right)^{1/2}} = \int \frac{du}{3u\left(u^2-1\right)^{1/2}} = \frac{1}{3}\sec^{-1}|u| + C = \frac{1}{3}\sec^{-1}\left|\frac{2}{3}x\right| + C.$$

43. Suggestion: Let $u = \frac{1}{5}x^3$.

46. Suggestion: Let $u = \sec x$.

49. Suggestion: Let $u = \ln x$.

52. $\displaystyle\int_0^1 \frac{x^3}{1+x^4}\,dx = \left[\frac{1}{4}\ln\left(1+x^4\right)\right]_0^1 = \frac{1}{4}\ln 2.$

58. Part (a): We begin with the identity
$$\tan(A+B) = \frac{\tan A + \tan B}{1 - \tan A \tan B}.$$
Let $x = \tan A$ and $y = \tan B$, and suppose that $xy < 1$. We will treat only the case in which x and y are both positive; the other cases are similar. In this case, the formula above shows that $0 < A + B < \pi/2$, so it is valid to apply the inverse tangent function to each side of the above identity to obtain
$$A + B = \arctan \frac{x+y}{1-xy},$$
and therefore
$$\arctan x + \arctan y = \arctan \frac{x+y}{1-xy}.$$

Part (b i): $\arctan \dfrac{(1/2)+(1/3)}{1-(1/6)} = \arctan(1) = \pi/4.$

Part (b ii): $\arctan \dfrac{(1/3)+(1/3)}{1-(1/9)} + \arctan(1/7) = \arctan(3/4) + \arctan(1/7)$
$$= \arctan \frac{(3/4)+(1/7)}{1-(3/28)} = \arctan \frac{25}{25} = \frac{\pi}{4}.$$

Part (b iii): $\arctan \dfrac{(120/119)-(1/239)}{1+(120/119)(1/239)} = \arctan \dfrac{28561/28441}{28561/28441} = \dfrac{\pi}{4}.$

Part (b iv): $2\arctan \dfrac{1}{5} = \arctan \dfrac{2/5}{1-(1/25)} = \arctan \dfrac{10}{24} = \arctan \dfrac{5}{12};$
$$4\arctan \frac{1}{5} = \arctan \frac{10/12}{1-(25/144)} = \arctan \frac{120}{119};$$
the rest of Part (b iv) follows from Part (b iii)

64. Let y be the height of the elevator (measured upward from ground level) and let θ be the angle that your line of sight to the elevator makes with the horizontal ($\theta > 0$ if you are looking up, $\theta < 0$ if down). You're to maximize $\dfrac{d\theta}{dt}$ given $\dfrac{dy}{dt} = -25$.

$$\tan\theta = \frac{y - 100}{50},$$

so

$$\theta = \tan^{-1}\left(\frac{y - 100}{50}\right).$$

Therefore

$$\frac{d\theta}{dt} = \frac{d\theta}{dy} \cdot \frac{dy}{dt} = -25\frac{1/50}{1 + ((y-100)/50)^2} = \frac{(-25)(50)}{2500 + (y - 100)^2}.$$

To find the value of y that maximizes $f(y) = d\theta/dt$, we need only minimize the last denominator above: $y = 100$. Answer: The elevator has maximum apparent speed when it's at eye level.

Section 8.3

1. $\displaystyle\lim_{x \to 1} \frac{x - 1}{x^2 - 1} = \lim_{x \to 1} \frac{1}{2x} = \frac{1}{2}$. Of course, l'Hôpital's rule is not necessary here, but it may be applied.

4. $\displaystyle\lim_{x \to 0} \frac{e^{3x} - 1}{x} = \lim_{x \to 0} \frac{3e^{3x}}{1} = 3.$

7. $\displaystyle\lim_{x \to 1} \frac{x - 1}{\sin x} = \frac{0}{\sin 1} = 0.$ Note that l'Hôpital's rule does not apply.

10. $\displaystyle\lim_{z \to \pi/2} \frac{1 + \cos 2z}{1 - \sin 2z} = \frac{1 + \cos\pi}{1 - \sin\pi} = \frac{0}{1} = 0.$

13. $\displaystyle\lim_{x \to \infty} \frac{\ln x}{x^{0.1}} = \lim_{x \to \infty} \frac{1/x}{(0.1)x^{-0.9}} = \lim_{x \to \infty} \frac{10}{x^{0.1}} = 0.$

16. $\displaystyle\lim_{t \to \infty} \frac{t^2 + 1}{t \ln t} = \lim_{t \to \infty} \frac{2t}{1 + \ln t} = \lim_{t \to \infty} \frac{2}{1/t} = +\infty.$

   It is also correct to say that this limit does not exist.

22. $\displaystyle\lim_{x \to 2} \frac{x^3 - 8}{x^4 - 16} = \frac{3}{8}.$

28. $\displaystyle\lim_{x \to \infty} \frac{\sqrt{x^3 + x}}{\sqrt{2x^3 - 4}} = \frac{1}{\sqrt{2}}.$

34. $\displaystyle\lim_{x \to 0} \frac{e^{3x} - e^{-3x}}{2x} = 3.$

40. $\displaystyle\lim_{x \to \infty} \frac{\arctan 2x}{\arctan 3x} = 1.$

46. $\displaystyle\lim_{x \to \pi/4} \frac{1 - \tan x}{4x - \pi} = -\frac{1}{2}.$

Section 8.4

1. $\displaystyle\lim_{x \to 0} x \cot x = \lim_{x \to 0} \frac{x \cos x}{\sin x} = (1)(\cos 0) = 1.$

4. Let $u = \sin x$. Then $\displaystyle\lim_{x \to 0+} (\sin x)(\ln \sin x) = \lim_{u \to 0+} u \ln u = \lim_{u \to 0+} \frac{\ln u}{1/u}$

$$= \lim_{u \to 0+} \frac{1/u}{-(1/u^2)} = \lim_{u \to 0+} (-u) = 0.$$

7. Let $u = 1/x$. Then $\lim\limits_{x \to \infty} x \left(e^{1/x} - 1 \right) = \lim\limits_{u \to 0+} \dfrac{e^u - 1}{u} = \lim\limits_{u \to 0+} \dfrac{e^u}{1} = 1.$

10. Replace $\cos 3x$ by $\cos^3 x - 3 \sin^2 x \cos x$; replace $\tan x$ by $(\sin x)/(\cos x)$.

Then $\lim\limits_{x \to \pi/2} (\tan x)(\cos 3x) = \lim\limits_{x \to \pi/2} (\sin x) \left(\cos^2 x - 3 \sin^2 x \right) = -3.$

16. $\lim\limits_{x \to \infty} \left(\sqrt{x+1} - \sqrt{x} \right) = \lim\limits_{x \to \infty} \dfrac{x + 1 - x}{\sqrt{x+1} + \sqrt{x}} = \lim\limits_{x \to \infty} \dfrac{1}{\sqrt{x+1} + \sqrt{x}} = 0.$

19. $\lim\limits_{x \to \infty} \left((x^3 + 2x + 5)^{1/3} - x \right) = \lim\limits_{x \to \infty} \dfrac{x^3 + 2x + 5 - x^3}{(x^3 + 2x + 5)^{2/3} + x(x^3 + 2x + 5)^{1/3} + x^2}$

$= \lim\limits_{x \to \infty} \dfrac{2x^{-1} + 5x^{-2}}{(1 + 2x^{-2} + 5x^{-3})^{2/3} + (1 + 2x^{-2} + 5x^{-3})^{1/3} + 1}$

$= \dfrac{0 + 0}{1 + 1 + 1} = 0.$

22. Let $y = \left(x + \dfrac{1}{x} \right)^x$. Then

$$\ln y = x \ln \left(1 + \dfrac{1}{x} \right) = x \ln \left(\dfrac{x+1}{x} \right) = x \left(\ln(x+1) - \ln x \right) = \dfrac{\ln(x+1) = \ln x}{1/x}.$$

Now apply l'Hôpital's rule to find that

$$\lim\limits_{x \to \infty} \ln y = \lim\limits_{x \to \infty} \dfrac{\ln(x+1) - \ln x}{1/x} = \lim\limits_{x \to \infty} \dfrac{x - (x+1)}{-(x+1)/x} = \lim\limits_{x \to \infty} \dfrac{x}{x+1} = 1.$$

Therefore $y \to e$ as $x \to +\infty$.

25. The limit of the natural logarithm of the expression in question is $-1/6$. A calculator estimate of this limit is particularly misleading.

28. $\lim\limits_{x \to 0+} (\sin x)^{\sec x} = 0.$

34. Let $Q = x^5 - 3x^4 + 17$. Then $Q^{1/5} - x = \dfrac{Q - x^5}{Q^{4/5} + Q^{3/5}x + Q^{2/5}x^2 + Q^{1/5}x^3 + x^4}.$

The numerator is just $-3x^4 + 17$. Now *carefully* divide each term in numerator and denominator by x^4; put this divisor within each radical in the denominator. Then let x increase without bound to see that the limit is $-3/5$.

Section 8.5

1. $f'(x) = 3 \sinh (3x - 2)$

4. $f'(x) = -2e^{2x} \operatorname{sech} e^{2x} \tanh e^{2x}$

10. $f'(x) = \dfrac{\operatorname{sech}^2 x}{1 + \tanh^2 x}$

16. $\displaystyle \int \cosh^2 3u \, du = \int \tfrac{1}{2}(1 + \cosh 6u) \, du = \tfrac{1}{2}(u + \tfrac{1}{6} \sinh 6u) + C = \tfrac{1}{2}u + \tfrac{1}{6} \sinh 3u \cosh 3u + C$

19. If necessary, let $u = 2x$.

22. $\displaystyle \int \sinh^4 x \, dx = \int \left(\sinh^2 x \right)^2 \, dx = \int \tfrac{1}{4} (\cosh 2x - 1)^2 \, dx = \tfrac{1}{4} \int \left(\cosh^2 2x - 2 \cosh 2x + 1 \right) \, dx$

$= \tfrac{1}{4} \int \left(\tfrac{1}{2}(\cosh 4x + 1) - 2 \cosh 2x + 1 \right) \, dx = \tfrac{1}{32} \sinh 4x - \tfrac{1}{4} \sinh 2x + \tfrac{3}{8}x + C$

28. $\displaystyle \int \dfrac{e^x + e^{-x}}{e^x - e^{-x}} \, dx = \ln |e^x - e^{-x}| + C_1 = -\ln 2 + \ln |e^x - e^{-x}| + C = \ln |\sinh x| + C$

34. $f'(x) = -\dfrac{e^x}{e^x (1 + e^{2x})^{1/2}} = -\dfrac{1}{(1 + e^{2x})^{1/2}}$

40. Let $u = \frac{2}{3}y$. Then $y = \frac{3}{2}u$, $dy = \frac{3}{2}\,du$, and $4y^2 - 9 = 9\left(u^2 - 1\right)$. So

$$\int \frac{dy}{\left(4y^2 - 9\right)^{1/2}} = \int \frac{(3/2)\,du}{2\left(u^2 - 1\right)^{1/2}} = \frac{1}{2}\int \frac{du}{\left(u^2 - 1\right)^{1/2}} = \frac{1}{2}\cosh^{-1} u + C = \frac{1}{2}\cosh^{-1}\left(\frac{2y}{3}\right) + C.$$

43. Let $u = \frac{3}{2}x$.

46. Let $u = x^2$. Then $du = 2x\,dx$, $x\,dx = \frac{1}{2}\,du$, and thus

$$\int \frac{x\,dx}{\left(x^4 - 1\right)^{1/2}} = \frac{1}{2}\int \frac{du}{\left(u^2 - 1\right)^{1/2}} = \frac{1}{2}\cosh^{-1} u + C = \frac{1}{2}\cosh^{-1}\left(x^2\right) + C.$$

49. Proof:

$$\begin{aligned}
\sinh x \cosh y + \cosh x \sinh y &= \tfrac{1}{4}\left(e^x - e^{-x}\right)\left(e^y + e^{-y}\right) + \tfrac{1}{4}\left(e^x + e^{-x}\right)\left(e^y - e^{-y}\right) \\
&= \tfrac{1}{4}\left(e^x e^y - e^{-x}e^y + e^x e^{-y} - e^{-x}e^{-y} + e^x e^y + e^{-x}e^y - e^x e^{-y} - e^{-x}e^{-y}\right) \\
&= \tfrac{1}{4}\left(2e^x e^y - 2e^{-x}e^{-y}\right) \\
&= \sinh(x + y).
\end{aligned}$$

52. If $x(t) = A\cosh kt + B\sinh kt$, then

$$x'(t) = kA\sinh kt + kB\cosh kt \text{ and}$$

$$x''(t) = k^2 A\cosh kt + k^2 B\sinh kt = k^2 x(t).$$

64. (a): We are to prove that

$$\coth^{-1} x = \frac{1}{2}\ln\left(\frac{x+1}{x-1}\right) \text{ if } |x| > 1.$$

By one of the formulas in the text, the left-hand side above has derivative

$$\frac{1}{1 - x^2}$$

The derivative of the right-hand side is

$$\left(\frac{1}{2}\right)\left(\frac{x-1}{x+1}\right)\left(\frac{x-1-x-1}{(x-1)^2}\right) = -\frac{1}{x^2-1} = \frac{1}{1-x^2}.$$

Therefore

$$\coth^{-1} x = \frac{1}{2}\ln\frac{x+1}{x-1} + C \text{ for } |x| > 1.$$

(b): If $u = \coth^{-1}(2)$, then $\coth u = 2$. So

$$2 = \frac{e^u + e^{-u}}{e^u - e^{-u}};$$

$$2e^u - 2e^{-u} = e^u + e^{-u};$$

$$e^u = 3e^{-u}; \quad e^{2u} = 3; \quad 2u = \ln 3.$$

Therefore $u = \frac{1}{2}\ln 3$. Now substitute $x = 2$ in the result of Part (a):

$$\frac{1}{2}\ln 3 = \frac{1}{2}\ln\frac{3}{1} + C.$$

So $C = 0$ here as well. This establishes the desired result.

Chapter 8 Miscellaneous

4. $g'(t) = \dfrac{e^t}{1 + e^{2t}}$

10. $f'(x) = -\dfrac{1}{x^2 + 1}$

16. $f'(x) = \dfrac{\sinh x}{\cosh x} = \tanh x$

22. $\displaystyle\int \frac{dx}{1 + 4x^2} = \tfrac{1}{2}\arctan 2x + C$

28. Let $u = \tfrac{2}{3}x$; $\displaystyle\int \frac{1}{9 + 4x^2}\, dx = \frac{3}{2}\int \frac{1}{9 + 9u^2}\, du = \tfrac{1}{6}\arctan u + C = \tfrac{1}{6}\arctan\left(\dfrac{2x}{3}\right) + C.$

31. Let $u = 2x$; $x = \tfrac{1}{2}u$ and $dx = \tfrac{1}{2}\, du$. Then

$$\int \frac{1}{x\sqrt{4x^2 - 1}}\, dx = \int \frac{du}{u\sqrt{u^2 - 1}} = \sec^{-1}|2x| + C.$$

34. $\displaystyle\int x^2 \cosh x^3\, dx = \tfrac{1}{3}\sinh x^3 + C.$

40. Let $u = x^2$; then $du = 2x\, dx$, and the integral becomes $\dfrac{1}{2}\displaystyle\int \frac{du}{\sqrt{u^2 + 1}} = \dfrac{1}{2}\sinh^{-1}\left(x^2\right) + C.$

43. $\displaystyle\lim_{x \to \pi} \frac{1 + \cos x}{(x - \pi)^2} = \lim_{x \to \pi} \frac{-\sin x}{2(x - \pi)} = \lim_{x \to \pi} \frac{-\cos x}{2} = \frac{1}{2}.$

46. $\displaystyle\lim_{x \to \infty} \frac{\ln(\ln x)}{\ln x} = \lim_{x \to \infty} \frac{\left(\dfrac{1}{x \ln x}\right)}{\left(\dfrac{1}{x}\right)} = \lim_{x \to \infty} \frac{1}{\ln x} = 0.$

49. $\displaystyle\lim_{x \to 0} \left(\frac{1}{x^2} - \frac{1}{1 - \cos x}\right) = \lim_{x \to 0} \frac{1 - \cos x - x^2}{x^2(1 - \cos x)} = \lim_{x \to 0} \frac{\sin x - 2x}{2x(1 - \cos x) + x^2 \sin x}$

$$= \lim_{x \to 0} \frac{\left(\dfrac{\sin x}{x} - 2\right)}{2(1 - \cos x) + x \sin x} = -\infty.$$

52. $\displaystyle\lim_{x \to \infty} \ln\left(x^{1/x}\right) = \lim_{x \to \infty} \frac{\ln x}{x} = 0.$ Therefore $\displaystyle\lim_{x \to \infty} x^{1/x} = 1.$

55. One of our most challenging problems. First let $u = 1/x$. It is then sufficient to evaluate

$$L = \lim_{u \to 0^+} \frac{(1 + u)^{1/u} - e}{u}.$$

Apply l'Hôpital's rule once:

$$L = \lim_{u \to 0^+} (1 + u)^{1/u} \left(\frac{u - (1 + u)\ln(1 + u)}{u^2(1 + u)}\right).$$

Now apply the product rule for limits!

$$L = e \lim_{u \to 0^+} \frac{u - (1 + u)\ln(1 + u)}{u^2(1 + u)}.$$

Finally apply l'Hôpital's rule twice to the limit that remains. The answer in the text follows without any more difficulties.

58. By the method of cylindrical shells,

$$V = \int_0^1 2\pi x \frac{1}{\sqrt{x^4 + 1}}\, dx.$$

Let $u = x^2$. Then $du = 2x\, dx$, and we obtain

$$V = \int_0^1 \frac{\pi}{\sqrt{u^2 + 1}}\, du = \pi\left(\sinh^{-1}(1) - \sinh^{-1}(0)\right) = \pi\sinh^{-1}(1) = \pi\ln(1 + \sqrt{2}) \approx 2.7689.$$

61. It's clear from the sketch on the right that the least positive solution is larger than $3\pi/2$ and smaller than 2π. So we use the iteration of Newton's method with initial estimate $x_0 = 5$. The formula is

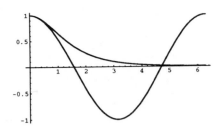

$$x \longleftarrow x - \frac{\cos x \cosh x - 1}{\cos x \sinh x - \sin x \cosh x}.$$

Here are the (rounded) results:

$$x_0 = 5.0$$
$$x_1 = 4.782556376$$
$$x_2 = 4.732575187$$
$$x_3 = 4.730047035$$
$$x_4 = 4.730040745 = x_5.$$

Answer: The least positive solution of $\cos x \cosh x = 1$ is approximately 4.730040745.

1. Let $u = 2 - 3x$ and apply Formula (1) of Fig. 9.1.

4. Let $u = 5 + 2t^2$. $\int \dfrac{5t}{5 + 2t^2}\, dt = \dfrac{5}{4} \int \dfrac{du}{u} = \dfrac{5}{4} \ln |u| + C = \dfrac{5}{4} \ln \left(5 + 2t^2\right) + C.$

7. Let $u = \sqrt{y}$.

10. Let $u = 4 + \cos 2x$. $\int \dfrac{\sin 2x}{4 + \cos 2x}\, dx = -\dfrac{1}{2} \int \dfrac{du}{u} = -\dfrac{1}{2} \ln |u| + C = -\dfrac{1}{2} \ln(4 + \cos 2x) + C.$

13. Let $u = \ln t$.

16. Let $u = 1 + e^{2x}$. $\int \dfrac{e^{2x}}{1 + e^{2x}}\, dx = \dfrac{1}{2} \int \dfrac{du}{u} = \dfrac{1}{2} \ln |u| + C = \dfrac{1}{2} \ln \left(1 + e^{2x}\right) + C$

19. Let $u = x^2$.

22. Let $u = 2t$. $\int \dfrac{1}{1 + 4t^2}\, dt = \dfrac{1}{2} \int \dfrac{du}{1 + u^2} = \dfrac{1}{2} \arctan u + C = \dfrac{1}{2} \arctan 2t + C.$

25. Let $v = 1 + \sqrt{x}$. $\int \dfrac{(1 + \sqrt{x})^4}{\sqrt{x}}\, dx = 2 \int v^4\, dv = \dfrac{2}{5} v^5 + C = \dfrac{2}{5} \left(1 + \sqrt{x}\right)^5 + C.$

28. Let $u = 1 + \sec 2x$. $\int \dfrac{\sec 2x \tan 2x}{(1 + \sec 2x)^{3/2}}\, dx = \int \dfrac{1}{2} u^{-3/2}\, du = -u^{-1/2} + C = -\dfrac{1}{\sqrt{1 + \sec 2x}} + C.$

31. $\displaystyle \int x^2 \sqrt{x - 2}\, dx = \int (u + 2)^2 u^{1/2}\, du = \int \left(u^{5/2} + 4u^{3/2} + 4u^{1/2}\right) du$

$$= \tfrac{2}{7} u^{7/2} + \tfrac{8}{5} u^{5/2} + \tfrac{8}{3} u^{3/2} + C = \tfrac{2}{7}(x - 2)^{7/2} + \tfrac{8}{5}(x - 2)^{5/2} + \tfrac{8}{3}(x - 2)^{3/2} + C$$
$$= (x - 2)^{3/2} \left(\tfrac{2}{7}(x - 2)^2 + \tfrac{8}{5}(x - 2) + \tfrac{8}{3}\right) + C$$
$$= \tfrac{1}{105}(x - 2)^{3/2} \left(30\left(x^2 - 4x + 4\right) + 168(x - 2) + 280\right) + C$$
$$= \tfrac{2}{105}(x - 2)^{3/2} \left(15x^2 + 24x + 32\right) + C.$$

34. $\displaystyle \int x \sqrt[3]{x - 1}\, dx = \int (u + 1) u^{1/3}\, du = \int \left(u^{4/3} + u^{1/3}\right) du$

$$= \dfrac{3}{7} u^{7/3} + \dfrac{3}{4} u^{4/3} + C = \left(3u^{4/3}\right)\left(\dfrac{4u + 7}{28}\right) + C$$
$$= \left(\dfrac{3(x - 1)^{4/3}}{28}\right)(4x + 3) + C = \tfrac{3}{28}(x - 1)^{1/3} \left(4x^2 - x - 3\right) + C.$$

37. With $a = 10$, $u = 3x$, and $dx = \tfrac{1}{3} du$, we get

$$\int \dfrac{1}{100 - 9x^2}\, dx = \dfrac{1}{3} \int \dfrac{1}{a^2 - u^2}\, du = \dfrac{1}{6a} \ln \left| \dfrac{u + a}{u - a} \right| + C = \dfrac{1}{60} \ln \left| \dfrac{3x + 10}{3x - 10} \right| + C.$$

40. With $a = 3$, $u = 4x$, and $dx = \tfrac{1}{4} du$, we obtain

$$\int \dfrac{dx}{\sqrt{16x^2 + 9}} = \dfrac{1}{4} \int \dfrac{du}{(u^2 + a^2)^{1/2}} = \dfrac{1}{4} \ln \left| u + (u^2 + a^2)^{1/2} \right| + C = \dfrac{1}{4} \ln \left| 4x + (16x^2 + 9)^{1/2} \right| + C.$$

43. With $a = 5$, $u = 4x$, and $dx = \tfrac{1}{4} du$, we get

$$\int x^2 \sqrt{25 - 16x^2}\, dx = \dfrac{1}{64} \int u^2 \left(a^2 - u^2\right)^{1/2}\, du$$
$$= \dfrac{1}{64} \left(\dfrac{u}{8}\left(2u^2 - a^2\right)\left(a^2 - u^2\right)^{1/2} + \dfrac{1}{8} a^4 \arcsin \dfrac{u}{a} \right) + C$$
$$= \dfrac{x}{128} \left(32x^2 - 25\right)\left(25 - 16x^2\right)^{1/2} + \dfrac{625}{512} \arcsin \dfrac{4x}{5} + C.$$

46. Let $u = \sin x$, $du = \cos x\, dx$. Use Formula (50) of the endpapers to obtain

$$\int \dfrac{\cos x}{(\sin^2 x)\sqrt{1 + \sin^2 x}}\, dx = \int \dfrac{1}{u^2 (1 + u^2)^{1/2}}\, du = -\dfrac{1}{u}\left(u^2 + 1\right)^{1/2} + C = -\dfrac{(1 + \sin^2 x)^{1/2}}{\sin x} + C$$

49. The substitution $u = \ln x$ leads to

$$\int \frac{(\ln x)^2}{x} \sqrt{1 + (\ln x)^2} \, dx = \int u^2 \left(1 + u^2\right)^{1/2} du \qquad \text{(which by Formula (48))}$$

$$= \frac{1}{8} \left((\ln x) \left(2(\ln x)^2 + 1\right) \left((\ln x)^2 + 1\right)^{1/2} - \ln \left| (\ln x) + \left((\ln x)^2 + 1\right)^{1/2} \right| \right) + C.$$

52. $\displaystyle \int \frac{dx}{x^2 + 4x + 5} = \int \frac{dx}{(x + 2)^2 + 1} = \tan^{-1}(x + 2) + C.$

Section 9.3

1. $\displaystyle \int \sin^2 2x \, dx = \int \frac{1 - \cos 4x}{2} \, dx = \frac{x}{2} - \frac{1}{8} \sin 4x + C.$

4. $\displaystyle \int \tan^2 \frac{x}{2} \, dx = \int \left(\sec^2 \frac{x}{2} - 1 \right) dx = \left(2 \tan \frac{x}{2} \right) - x + C.$

7. $\displaystyle \int \sec 3x \, dx = \frac{1}{3} \ln |\sec 3x + \tan 3x| + C.$

10. $\displaystyle \int \sin^2 x \cot^2 x \, dx = \int \cos^2 x \, dx = \int \frac{1 + \cos 2x}{2} \, dx = \frac{1}{2}x + \frac{1}{4} \sin 2x + C.$

13. $\displaystyle \int \sin^2 \theta \cos^3 \theta \, d\theta = \int \left(\sin^2 \theta\right) \left(1 - \sin^2 \theta\right) (\cos \theta) \, d\theta$

$$= \int \left(\sin^2 \theta \cos \theta - \sin^4 \theta \cos \theta\right) d\theta = \frac{1}{3} \sin^3 \theta - \frac{1}{5} \sin^5 \theta + C.$$

16. $\displaystyle \int \frac{\sin t}{\cos^3 t} \, dt = \frac{1}{2 \cos^2 t} + C.$

19. $\displaystyle \int \sin^5 2z \cos^2 2z \, dz = \int (\sin 2z) \left(1 - \cos^2 2z\right)^2 \cos^2 2z \, dz$

$$= \int \left(\cos^6 2z \sin 2z - 2 \cos^4 2z \sin 2z + \cos^2 2z \sin 2z\right) dz$$

$$= -\frac{1}{14} \cos^7 2z + \frac{1}{5} \cos^5 2z - \frac{1}{6} \cos^3 2z + C.$$

22. $\cos^6 u = \left(\cos^2 u\right)^3 = \frac{1}{8} \left(1 + \cos 2u\right)^3 = \frac{1}{8} \left(1 + 3 \cos 2u + 3 \cos^2 2u + \cos^3 2u\right)$

$$= \frac{1}{8} \left(1 + 3 \cos 2u + \frac{3}{2} \left(1 + \cos 4u\right) + \left(1 - \sin^2 2u\right) (\cos 2u)\right).$$

Thus the antiderivative—simplified—is $\frac{5}{16}\theta + \frac{1}{16} \sin 8\theta + \frac{3}{256} \sin 16\theta - \frac{1}{192} \sin^3 8\theta + C.$

25. $\displaystyle \int \cot^3 2x \, dx = \int (\cot 2x) \left(\csc^2 2x - 1\right) dx.$

28. $\displaystyle \int \cot^3 x \csc^2 x \, dx = -\frac{1}{4} \cot^4 x + C.$

31. $\displaystyle \int \frac{\tan^3 \theta}{\sec^4 \theta} \, d\theta = \int (\sec \theta)^{-4} \left(\sec^2 \theta - 1\right) (\tan \theta) \, d\theta$

$$= \int \left((\sec \theta)^{-3} \sec \theta \tan \theta - (\sec \theta)^{-5} \sec \theta \tan \theta\right) d\theta.$$

34. $\displaystyle \int \frac{1}{\cos^4 2x} \, dx = \int \sec^4 2x \, dx = \int \left(\tan^2 2x + 1\right) \sec^2 2x \, dx = \frac{1}{6} \tan^3 2x + \frac{1}{2} \tan 2x + C.$

37. $\displaystyle \int \cos^3 5t \, dt = \int \left(1 - \sin^2 5t\right) \cos 5t \, dt = \int \left(\cos 5t - \sin^2 5t \cos 5t\right) dt.$

40. $\displaystyle \int \tan^2 2t \sec^4 2t \, dt = \int \left(\tan^2 2t\right) \left(\tan^2 2t + 1\right) \sec^2 2t \, dt = \int \left(\tan^4 2t \sec^2 2t + \tan^2 2t \sec^2 2t\right) dt$

$$= \frac{1}{10} \tan^5 2t + \frac{1}{6} \tan^3 2t + C.$$

43. $\displaystyle \frac{\tan x}{\sec x} = \sin x$ and $\displaystyle \frac{\sin x}{\sec x} = \sin x \cos x.$

46. $\displaystyle \int \cot^n x\, dx = \int \left(\cot^{n-2} x \right) \left(\csc^2 x - 1 \right)\, dx$

$$= \int \cot^{n-2} x\, \csc^2 x\, dx - \int \cot^{n-2} x\, dx$$

$$= -\frac{1}{n-1} \cot^{n-1} x - \int \cot^{n-2} x\, dx.$$

49. $\sin 3x \cos 5x = \frac{1}{2}(\sin 8x - \sin 2x).$

52. (a) $\displaystyle \int_0^{2\pi} \sin mx \sin nx\, dx = \frac{1}{2} \int_0^{2\pi} \left(\cos(n-m)x - \cos(n+m)x \right)\, dx$

$$= \frac{1}{2} \left[\frac{\sin(n-m)x}{n-m} - \frac{\sin(n+m)x}{n+m} \right]_0^{2\pi} = 0.$$

The derivations for Parts (b) and (c) are similar.

58. Volume: $\displaystyle V = \int_0^{\pi/4} \pi \sec^2 x\, dx = \pi \left[\tan x \right]_0^{\pi/4} = \pi.$

Section 9.4

1. $\begin{bmatrix} u = x & dv = e^{2x}\, dx \\ du = dx & v = \frac{1}{2}e^{2x} \end{bmatrix}$ $\displaystyle \int x e^{2x}\, dx = \frac{1}{2} x e^{2x} - \int \frac{1}{2} e^{2x}\, dx = \frac{1}{2} x e^{2x} - \frac{1}{4} e^{2x} + C.$

4. $\begin{bmatrix} u = t^2 & dv = \sin t\, dt \\ du = 2t\, dt & v = -\cos t \end{bmatrix}$ $\displaystyle \int t^2 \sin t\, dt = -t^2 \cos t + 2 \int t \cos t\, dt.$

7. $\begin{bmatrix} u = \ln x & dv = x^3\, dx \\ du = \dfrac{1}{x}\, dx & v = \frac{1}{4} x^4 \end{bmatrix}$ $\displaystyle \int x^3 \ln x\, dx = \frac{1}{4} x^4 \ln x - \frac{1}{16} x^4 + C.$

10. $\begin{bmatrix} u = \ln x & dv = \dfrac{1}{x^2}\, dx \\ du = \dfrac{1}{x}\, dx & v = -\dfrac{1}{x} \end{bmatrix}$ $\displaystyle \int \frac{\ln x}{x^2}\, dx = -\frac{1}{x} \ln x + \int \frac{1}{x^2}\, dx = -\frac{1}{x} \ln x - \frac{1}{x} + C.$

13. Choose $u = (\ln t)^2$ and $dv = dt$. The necessity of integrating $\ln t$ then arises; a second integration by parts is necessary, and works well with the choices $u = \ln t$, $dv = dt$.

16. $\begin{bmatrix} u = x^2 & dv = x \left(1 - x^2 \right)^{1/2}\, dx \\ du = 2x\, dx & v = -\frac{1}{3} \left(1 - x^2 \right)^{3/2} \end{bmatrix}$ $\displaystyle \int x^3 \sqrt{1 - x^2}\, dx = -\frac{1}{3} x^2 \left(1 - x^2 \right)^{3/2} - \frac{2}{15} \left(1 - x^2 \right)^{5/2} + C.$

19. Choose $u = \csc \theta$, $dv = \csc^2 \theta\, d\theta$. Replace $\cot^2 \theta$ by $\csc^2 \theta - z$ in the resulting integral to obtain

$$I = -\csc \theta \cot \theta + \int \csc \theta\, d\theta - I$$

where I is the original integral. Solve for I and use Formula 15 of the endpapers to obtain the answer.

22. Choose $u = \ln \left(1 + x^2 \right)$, $dv = dx$. Then

$$\int \ln \left(1 + x^2 \right)\, dx = x \ln \left(1 + x^2 \right) - 2 \int \frac{x^2}{x^2 + 1}\, dx$$

$$= x \ln \left(1 + x^2 \right) - 2 \int \frac{x^2 + 1}{x^2 + 1}\, dx + 2 \int \frac{dx}{x^2 + 1}$$

$$= x \ln \left(1 + x^2 \right) - 2x + 2 \tan^{-1} x + C.$$

25. Choose $u = \tan^{-1} \sqrt{x}$ and $dv = dx$. To simplify the resulting computations, choose $v = x + 1$.

28. Choose $u = \tan^{-1} x$, $dv = x\, dx$. Then $du = \dfrac{dx}{1+x^2}$; we choose $v = \frac{1}{2}x^2 + \frac{1}{2} = \frac{1}{2}\left(x^2 + 1\right)$ for the purpose of providing a fortuitous cancellation:

$$\int x \tan^{-1} x\, dx = \frac{1}{2}\left(x^2 + 1\right)\tan^{-1} x - \frac{1}{2}\int \frac{x^2 + 1}{1+x^2}\, dx = \frac{1}{2}\left(x^2 + 1\right)\tan^{-1} x - \frac{1}{2}x + C.$$

This sly trick—a clever choice for v—is rarely useful. Try it with the integral of $\ln(x+1)$.

31. Choose $u = \ln x$ and $dv = x^{-3/2}\, dx$. Then

$$\int \frac{\ln x}{x^{3/2}}\, dx = -2x^{-1/2}\ln x + 2\int x^{-3/2}\, dx$$

$$= -2x^{-1/2}\ln x - 4x^{-1/2} + C$$

$$= -\frac{2}{\sqrt{x}}\left(2 + \ln x\right) + C.$$

34. First method:

$$\int e^x \cosh x\, dx = \frac{1}{2}\int \left(e^{2x} + 1\right)\, dx = \tfrac{1}{4}e^{2x} + \tfrac{1}{2}x + C_1$$

$$= \frac{1}{4}\left(e^{2x} + 1\right) - \frac{1}{4} + \frac{1}{2}x + C_1 = \frac{1}{4}e^x\left(e^x + e^{-x}\right) + \frac{1}{2}x + C$$

$$= \frac{1}{2}e^x \cosh x + \frac{1}{2}x + C.$$

Second method: Presented because no integration by parts is used in the first method, although what follows is somewhat artificial.

$$\begin{bmatrix} u = e^x & dv = \cosh x\, dx \\ du = e^x\, dx & v = \sinh x \end{bmatrix} \qquad J = \int e^x \cosh x\, dx = e^x \sinh x - \int e^x \sinh x\, dx.$$

Now $e^x \sinh x = \frac{1}{2}\left(e^{2x} - 1\right) = \frac{1}{2}\left(e^{2x} + 1\right) - 1 = e^x \cosh x - 1$. Therefore

$$J = e^x \sinh x - J + \int 1\, dx; \text{ it follows that } \int e^x \cosh x\, dx = \tfrac{1}{2}e^x \sinh x + \tfrac{1}{2}x + C.$$

37. $V = \displaystyle\int_1^e \pi(\ln x)^2\, dx$. Now apply the solution of Problem 13.

40. Given: Constants A and B, neither zero, $A \neq B$, and $J = \displaystyle\int \sin Ax \cos Bx\, dx$.

Let $u = \sin Ax$ and $dv = \cos Bx\, dx$. Result:

$$J = \frac{1}{B}\sin Ax \sin Bx + \frac{A}{B}\int \cos Ax \sin Bx + C.$$

In the second integral, let $u = \cos Ax$ and $dv = \sin Bx\, dx$ (the other choice doesn't work). You will find that

$$J = \frac{1}{B}\sin Ax \sin Bx + \frac{A}{B^2}\cos Ax \cos Bx + \frac{A^2}{B^2}J.$$

Now solve for J to obtain

$$J = \frac{B}{B^2 - A^2}\sin Ax \sin Bx + \frac{A}{B^2 - A^2}\cos Ax \cos Bx + C.$$

In particular, we get the integral in Problem 40 by choosing $A = 3$ and $B = 1$, thus obtaining

$$\int \sin 3x \cos x\, dx = -\tfrac{1}{8}\sin 3x \sin x - \tfrac{3}{8}\cos 3x \cos x + C.$$

See Problems 49–52 of Section 9.3 for a "better" way, which yields the antiderivative in the alternative form $-\frac{1}{8}\cos 4x - \frac{1}{4}\cos 2x + C$.

43. Let $u = (\ln x)^n$, $dv = dx$.

46. $\left[\begin{array}{ll} u = \cos^{n-1} x & dv = \cos x\, dx \\ du = -(n-1)\cos^{n-2} x \sin x\, dx & v = \sin x \end{array}\right]$

$$I_n = \int \cos^n x\, dx$$

$$= \cos^{n-1} x \sin x + \int (n-1)\cos^{n-2} x \sin^2 x\, dx$$

$$= \cos^{n-1} x \sin x + (n-1)I_{n-2} - (n-1)I_n$$

(upon replacement of $\sin^2 x$ by $1 - \cos^2 x$).

Therefore, $nI_n = \cos^{n-1} x \sin x + (n-1)I_{n-2}$; that is,

$$\int \cos^n x\, dx = \frac{1}{n} \cos^{n-1} x \sin x + \frac{n-1}{n} \int \cos^{n-2} x\, dx.$$

49. $\int (\ln x)^3\, dx = x(\ln x)^3 - 3\int x(\ln x)^2\, dx - 2\int x \ln x\, dx - \int 1\, dx.$

So $\int_1^e (\ln x)^3\, dx = \left[x(\ln x)^3 - 3x(\ln x)^2 + 6x(\ln x) - 6x\right]_1^e = e - 3e + 6e - 6e + 6 = 6 - 2e \approx 0.563436.$

Section 9.5

1. $\int \dfrac{x^2}{x+1}\, dx = \int \left(x - 1 + \dfrac{1}{x+1}\right) dx = \tfrac{1}{2}x^2 - x + \ln|x+1| + C.$

4. $\int \dfrac{x}{x^2 + 4x}\, dx = \int \dfrac{1}{x+4}\, dx = \ln|x+4| + C.$

7. $\dfrac{1}{x^3 + 4x} = \dfrac{1}{x(x^2+4)} = \dfrac{A}{x} + \dfrac{Bx+C}{x^2+4}$ yields $A = \tfrac{1}{4}$, $B = -\tfrac{1}{4}$, and $C = 0$.

So $\int \dfrac{1}{x^3 + 4x}\, dx = \tfrac{1}{4}\ln|x| - \tfrac{1}{8}\ln(x^2+4) + C.$

10. $\dfrac{1}{(x^2+1)(x^2+4)} = \dfrac{Ax+B}{x^2+1} + \dfrac{Cx+D}{x^2+4}$ yields $A = C = 0$, $B = \tfrac{1}{3}$, and $D = -\tfrac{1}{3}$.

Thus $\int \dfrac{1}{(x^2+1)(x^2+4)}\, dx = \tfrac{1}{3}\arctan x - \tfrac{1}{6}\arctan\left(\dfrac{x}{2}\right) + C.$

13. Rewrite the integrand as $1 - \dfrac{1}{(x+1)^2}.$

16. After division:

$$\int \frac{x^4}{x^2 + 4x + 4}\, dx = \int \left(x^2 - 4x + 12 - (16)\frac{2x+3}{(x+2)^2}\right) dx$$

$$= \tfrac{1}{3}x^3 - 2x^2 + 12x - 16\int \frac{2x + 4 - 1}{(x+2)^2}\, dx$$

$$= \tfrac{1}{3}x^3 - 2x^2 + 12x - 16\int \left(\frac{2}{x+2} - \frac{1}{(x+2)^2}\right) dx$$

$$= \tfrac{1}{3}x^3 - 2x^2 + 12x - \frac{16}{x+2} - 32\ln|x+2| + C.$$

19. $\dfrac{x^2+1}{x^3 + 2x^2 + x} = \dfrac{1}{x} + \dfrac{0}{x+1} - \dfrac{2}{(x+1)^2}.$

22. $\int \dfrac{2x^2 + 3}{x^4 - 2x^2 + 1}\, dx = \int \left(-\dfrac{1}{4}\left(\dfrac{1}{x-1}\right) + \dfrac{5}{4}\left(\dfrac{1}{(x-1)^2}\right) + \dfrac{1}{4}\left(\dfrac{1}{x+1}\right) + \dfrac{5}{4}\left(\dfrac{1}{(x+1)^2}\right)\right) dx$

$$= \frac{1}{4}\ln\left|\frac{x+1}{x-1}\right| - \frac{5x}{2(x^2-1)} + C.$$

25. $\dfrac{1}{x^3 + x} = \dfrac{A}{x} + \dfrac{Bx + C}{x^2 + 1}$ leads to $A = 1$, $B = -1$, $C = 0$.

28. $\dfrac{4x^4 + x + 1}{x^5 + x^4} = \dfrac{A}{x} + \dfrac{B}{x^2} + \dfrac{C}{x^3} + \dfrac{D}{x^4} + \dfrac{E}{x + 1}$ yields $A = B = C = 0$, $D = 1$, and $E = 4$.

So $\displaystyle\int \dfrac{4x^4 + x + 1}{x^5 + x^4}\, dx = \int \left(\dfrac{1}{x^4} + \dfrac{4}{x + 1}\right) dx = -\tfrac{1}{3}x^{-3} + 4\ln|x + 1| + C.$

31. $\dfrac{x^2 - 10}{2x^4 + 9x^2 + 4} = \dfrac{Ax + B}{2x^2 + 1} + \dfrac{Cx + D}{x^2 + 4}$ yields $A = C = 0$, $B = -3$, and $D = 2$.

34. $\displaystyle\int \dfrac{x^2 + 4}{(x^2 + 1)^2 (x^2 + 2)}\, dx = \int \left(-\dfrac{2}{x^2 + 1} + \dfrac{3}{(x^2 + 1)^2} + \dfrac{2}{x^2 + 2}\right) dx.$

The first term in the integrand presents no problem, and by Formula 17 of the endpapers the third yields

$$\sqrt{2}\arctan\left(\tfrac{1}{2}x\sqrt{2}\right) + C_3.$$

For the second term, we use the substitution $x = \tan z$. Then

$$\int \dfrac{3}{(x^2 + 1)^2}\, dx = 3\int \dfrac{\sec^2 z}{\sec^4 z}\, dz = 3\int \tfrac{1}{2}(1 + \cos 2z)\, dz$$

$$= \tfrac{3}{2}(z + \sin z \cos z) + C_2 = \tfrac{3}{2}\left(\tan^{-1} x + \dfrac{x}{1 + x^2}\right) + C_2.$$

When we assemble this work, we find:

$$\int \dfrac{x^2 + 4}{(x^2 + 1)^2 (x^2 + 2)}\, dx = \dfrac{3x}{2(1 + x^2)} - \dfrac{1}{2}\tan^{-1} x + \sqrt{2}\tan^{-1}\left(\tfrac{1}{2}x\sqrt{2}\right) + C.$$

37. Let $x = e^{2t}$. Then $dx = 2e^{2t}\, dt$; $e^{4t}\, dt = \tfrac{1}{2}x\, dx$.

$$\int \dfrac{e^{4t}}{(e^{2t} - 1)^3}\, dt = \dfrac{1}{2}\int \dfrac{x}{(x - 1)^3}\, dx = \dfrac{1}{2}\int \left(\dfrac{1}{(x - 1)^2} + \dfrac{1}{(x - 1)^3}\right) dx$$

$$= -\dfrac{1}{4}\left(\dfrac{2}{x - 1} + \dfrac{1}{(x - 1)^2}\right) + C = -\dfrac{2x - 1}{4(x - 1)^2} + C = \dfrac{1 - 2e^{2t}}{4(e^{2t} - 1)^2} + C.$$

40. Let V be the volume of the solid. Then

$$V = \int_0^1 \pi y^2\, dx = \pi \int_0^1 x^2 \left(\dfrac{1 - x}{1 + x}\right) dx$$

$$= \pi \int_0^1 \left(-x^2 + 2x - 2 + \dfrac{2}{x + 1}\right) dx$$

$$= \pi \left[-\tfrac{1}{3}x^3 + x^2 - 2x + 2\ln|x + 1|\right]_0^1$$

$$= \pi \left(-\tfrac{1}{3} + 1 - 2 + 2\ln 2\right) = \tfrac{2}{3}\pi(3\ln 2 - 2) \approx 0.166382.$$

46. $x(t) = \dfrac{3e^{12t} - 3}{2e^{12t} + 2}$

52. $P(t) = 200$ when $t = 50\ln(1.125) \approx 5.89$ (months).

Section 9.6

1. Let $x = 4\sin u$. Then $16 - x^2 = 16\cos^2 u$ and $dx = 4\cos u\, du$.

$$\int \dfrac{1}{\sqrt{16 - x^2}}\, dx = \int \dfrac{4\cos u}{4\cos u}\, du = u + C = \arcsin\dfrac{x}{4} + C$$

4. Let $x = 5 \sec u$. Then $x^2 - 25 = 25\left(\sec^2 u - 1\right) = 25\tan^2 u$ and $dx = 5\sec u \tan u \, du$.

$$\int \frac{1}{x^2\sqrt{x^2-25}}\,dx = \int \frac{5\sec u \tan u}{\left(25\sec^2 u\right)\left(5\tan u\right)}\,du = \tfrac{1}{25}\int \cos u \, du = \tfrac{1}{25}\sin u + C = \frac{\sqrt{x^2-25}}{25x} + C$$

7. Let $x = \tfrac{3}{4}\sin u$. Then $9 - 16x^2 = 9\cos^2 u$ and $dx = \tfrac{3}{4}\cos u \, du$.

$$\int \frac{1}{\left(9-16x^2\right)^{3/2}}\,dx = \int \frac{1}{27\cos^3 u}\left(\tfrac{3}{4}\cos u\right)\,du = \tfrac{1}{36}\int \sec^2 u \, du = \tfrac{1}{36}\tan u + C = \frac{x}{9\sqrt{9-16x^2}} + C$$

10. Let $x = 2\sin u$. Then $dx = 2\cos u \, du$ and $\sqrt{4-x^2} = 2\cos u$.

$$\int x^3\sqrt{4-x^2}\,dx = \int 32\sin^3 u \, \cos^2 u \, du = 32\int (\sin u)\left(\cos^2 u - \cos^4 u\right)\,du$$

$$= 32\left(\tfrac{1}{5}\cos^5 u - \tfrac{1}{3}\cos^3 u\right) + C = \tfrac{32}{15}\left(3\left(\frac{\sqrt{4-x^2}}{2}\right)^5 - 5\left(\frac{\sqrt{4-x^2}}{2}\right)^3\right) + C$$

$$= \tfrac{4}{15}\left(\tfrac{3}{4}\left(4-x^2\right)\left(4-x^2\right)^{3/2} - 5\sqrt{4-x^2}^{\,3/2}\right) + C$$

$$= -\tfrac{1}{15}\left(4-x^2\right)^{3/2}\left(3x^2 + 8\right) + C$$

13. Let $x = \tfrac{1}{2}\sin\theta$. Then $1 - 4x^2 = \cos^2\theta$, and the integrand becomes $\csc\theta - \sin\theta$.

16. Let $x = \tfrac{1}{2}\tan z$. Then $2x = \tan z$, $1 + 4x^2 = \sec^2 z$, and $dx = \tfrac{1}{2}\sec^2 z \, dz$.

$$\int \sqrt{1+4x^2}\,dx = \int \tfrac{1}{2}(\sec z)\left(\sec^2 z\right)\,dz = \tfrac{1}{2}\int \sec^3 z \, dz$$

Now do a moderately difficult integration by parts, or apply Formula 28 of the endpapers, to obtain

$$\int \sqrt{1+4x^2}\,dx = \tfrac{1}{4}\left(\sec z \tan z + \ln\left|\sec z + \tan z\right|\right) + C$$

$$= \tfrac{1}{4}\left(2x\sqrt{1+4x^2} + \ln\left|2x + \sqrt{1+4x^2}\,\right|\right) + C$$

19. Use the substitution $x = \tan u$ to transform the integrand into $\tan^2 u \sec u = \sec^3 u - \sec u$, then apply Formulas 14 and 28 of the endpapers.

22. Let $x = \sin z$. Then $\sqrt{1-x^2} = \cos z$ and $dx = \cos z \, dz$.

$$\int \left(1-x^2\right)^{3/2}\,dx = \int \cos^4 z \, dz$$

$$= \tfrac{1}{4}\int \left(1 + \cos 2z\right)^2\,dz = \tfrac{1}{4}\int \left(1 + 2\cos 2z + \tfrac{1}{2}(1 + \cos 4z)\right)\,dz$$

$$= \tfrac{1}{4}\left(\tfrac{3}{2}z + \sin 2z + \tfrac{1}{8}\sin 4z\right) + C = \tfrac{3}{8}z + \tfrac{1}{2}\sin z \, \cos z + \tfrac{1}{16}\sin 2z \, \cos 2z + C$$

$$= \tfrac{3}{8}z + \tfrac{1}{2}\sin z \, \cos z + \tfrac{1}{8}(\sin z \, \cos z)\left(\cos^2 z - \sin^2 z\right) + C$$

$$= \tfrac{3}{8}z + \tfrac{1}{2}\sin z \, \cos z + \tfrac{1}{8}\sin z \, \cos^3 z - \tfrac{1}{8}\sin^3 z \, \cos z + C$$

$$= \tfrac{3}{8}\arcsin x + \tfrac{1}{2}x\sqrt{1-x^2} + \tfrac{1}{8}x\left(1-x^2\right)^{3/2} - \tfrac{1}{8}x^3\sqrt{1-x^2} + C$$

$$= \tfrac{1}{8}\left(3\arcsin x + x\left(5 - 2x^2\right)\sqrt{1-x^2}\right) + C$$

25. The substitution $x = 2\sin u$ transforms the integral into $\tfrac{1}{32}\int \sec^5 u \, du$. Application of Formula 37 of the endpapers transforms this integral into

$$\tfrac{1}{32}\int \tfrac{1}{4}\sec^3 u \, \tan u \, du + \tfrac{3}{4}\int \sec^3 u \, du.$$

Then Formula 28 from the endpapers yields the antiderivative

$$\tfrac{1}{128} \sec^3 u \tan u + \tfrac{3}{256} \sec u \tan u + \tfrac{3}{256} \ln|\sec u + \tan u| + C.$$

Finally make the replacements $\sec u = \dfrac{2}{\sqrt{4-x^2}}$ and $\tan u = \dfrac{x}{\sqrt{4-x^2}}$ to obtain the final answer.

28. Use the substitution $x = \tfrac{3}{4} \tan u$ to obtain $\tfrac{81}{4} \displaystyle\int \sec^5 u \, du$. With the aid of Formula 37 of the endpapers, we next obtain

$$\int (9 + 16x^2)^{3/2} \, dx = \tfrac{81}{4} \left(\tfrac{1}{4} \sec^3 u \tan u + \tfrac{3}{8} \sec u \tan u + \tfrac{3}{8} \ln|\sec u + \tan u| \right) + C$$

$$= \tfrac{1}{4} x \left(9 + 16x^2 \right)^{3/2} + \tfrac{27}{8} x \left(9 + 16x^2 \right)^{1/2} + \tfrac{243}{32} \ln \left| 4x + \sqrt{9 + 16x^2} \right| + C$$

(We have allowed the constant $-\tfrac{243}{32} \ln 3$ to be "absorbed" by the constant C.)

31. Let $x = \sec u$. Then $x^2 - 1 = \tan^2 x$ and $dx = \sec u \tan u \, du$.

$$\int x^2 \sqrt{x^2 - 1} \, dx = \int \sec^3 u \tan^2 u \, du = \int \left(\sec^5 u - \sec^3 u \right) \, du$$

$$= \tfrac{1}{4} \sec^3 u \tan u - \tfrac{1}{8} \sec u \tan u - \tfrac{1}{8} \ln|\sec u + \tan u| + C$$

—with the aid of Formula 37 of the endpapers. Then replace $\tan u$ by $\sqrt{x^2 - 1}$ and $\sec u$ by x to obtain the final version of the answer.

34. Let $x = \tfrac{3}{2} \sec u$. Then $\sqrt{4x^2 - 9} = \sqrt{9 \sec^2 u - 9} = 3 \tan u$ and $dx = \tfrac{3}{2} \sec u \tan u \, du$.

$$\int \frac{1}{x^2 \sqrt{4x^2 - 9}} \, dx = \tfrac{2}{9} \int \cos u \, du = \tfrac{2}{9} \sin u + C = \frac{\sqrt{4x^2 - 9}}{9x} + C$$

37. The substitution $x = 5 \sinh u$ yields

$$\int \frac{dx}{\sqrt{25 + x^2}} \, dx = \int \frac{5 \cosh u}{5 \cosh u} \, du = \int 1 \, du = u + C = \sinh^{-1} \left(\frac{x}{5} \right) + C.$$

40. Let $x = \tfrac{1}{3} \sinh u$. Then $9x^2 = \sinh^2 u$, $\sqrt{1 + 9x^2} = \cosh u$, and $dx = \tfrac{1}{3} \cosh u \, du$.

$$\int \frac{dx}{\sqrt{1 + 9x^2}} \, dx = \tfrac{1}{3} \int \frac{\cosh u}{\cosh u} \, du = \tfrac{1}{3} u + C = \tfrac{1}{3} \sinh^{-1}(3x) + C$$

43. $A = \displaystyle\int_0^1 2\pi x^2 \sqrt{1 + 4x^2} \, dx$. Let $x = \tfrac{1}{2} \tan u$. This yields the transformation

$$I = \int x^2 \sqrt{1 + 4x^2} \, dx = \tfrac{1}{8} \int \sec^3 u \tan^2 u \, du = \tfrac{1}{8} \int \left(\sec^5 u - \sec^3 u \right) \, du.$$

Apply Formula 37 of the endpapers with $n = 5$, then Formula 28:

$$I = \tfrac{1}{32} \left(\sec^3 u \tan u - \tfrac{1}{2} \sec u \tan u - \tfrac{1}{2} \ln|\sec u + \tan u| \right) + C$$

$$= \tfrac{1}{64} \left(4x \left(1 + 4x^2 \right)^{3/2} - 2x \left(1 + 4x^2 \right)^{1/2} - \ln \left| 2x + \sqrt{1 + 4x^2} \right| \right) + C.$$

Now to obtain A, substitute $x = 1$, subtract the value when $x = 0$, and finally multiply by 2π; the result is that

$$A = \frac{\pi}{32} \left(18\sqrt{5} - \ln \left(2 + \sqrt{5} \right) \right) \approx 3.80973.$$

46. $A = \displaystyle\int_1^2 2\pi x \frac{\sqrt{x^2 + 1}}{x} \, dx = 2\pi \int_1^2 \sqrt{x^2 + 1} \, dx$. The substitution $x = \tan u$ transforms the antidifferentiation problem into

$$2\pi \int \sec^3 z \, dz = \pi \left(\sec z \tan z + \ln|\sec z + \tan z| \right) + C = \pi \left(x\sqrt{x^2 + 1} + \ln \left| x + \sqrt{x^2 + 1} \right| \right) + C.$$

Substitution of the limits $x =$ and $x = 2$ yields the answer:

$$A = \pi\left(2\sqrt{5} - \sqrt{2} + \ln\left(2 + \sqrt{5}\right) - \ln\left(1 + \sqrt{2}\right)\right) \approx 11.37314434.$$

49. $A = 4\pi \displaystyle\int_0^{\pi/2} (\sin x)\sqrt{1 + \cos^2 x}\, dx$. With $u = \cos x$ and $du = -\sin x\, dx$, we obtain

$$A = 4\pi \int_0^1 \sqrt{1 + u^2}\, du.$$

To find the antiderivative, we let $u = \sinh z$, $du = \cosh z\, dz$. Then we obtain

$$A = 4\pi \int_{u=0}^{u=1} \cosh^2 z\, dz = \Big[2\pi(z + \sinh z \,\cosh z)\Big]_{u=0}^{u=1}$$

$$= 2\pi\left[\sinh^{-1} u + u\sqrt{1 + u^2}\right]_0^1 = 2\pi\left(\sinh^{-1}(1) + \sqrt{2}\right)$$

$$= 2\pi\left(\sqrt{2} + \ln\left(1 + \sqrt{2}\right)\right) \approx 14.4236.$$

52. You should find that the arc length element is $ds = \dfrac{\sqrt{x}}{\sqrt{x-1}}\, dx$. After using the suggested substitution, it turns out that the antiderivative is

$$\left(\sqrt{x}\,\sqrt{x-1} + \ln\left(\sqrt{x} + \sqrt{x-1}\right)\right) + C,$$

and that the value of the definite integral is approximately equal to 3.620184.

Section 9.7

1. $\displaystyle\int \frac{1}{x^2 + 4x + 5}\, dx = \int \frac{1}{(x+2)^2 + 1}\, dx = \arctan(x+2) + C.$

4. Let $u = x + 2$: $x = u - 2$ and $dx = du$. Thus

$$\int \frac{x+1}{(x^2 + 4x + 5)^2}\, dx = \int \frac{u-1}{(u^2 + 1)^2}\, du.$$

Next let $u = \tan\theta$. Then $du = \sec^2\theta\, d\theta$, and

$$\int \frac{x+1}{(x^2 + 4x + 5)^2}\, dx = \int \frac{\tan\theta - 1}{\sec^2\theta}\, d\theta = \int \left(\sin\theta\,\cos\theta - \cos^2\theta\right)\, d\theta$$

$$= \tfrac{1}{2}\sin^2\theta - \tfrac{1}{2}\theta - \tfrac{1}{2}\sin\theta\,\cos\theta + C$$

$$= \frac{u^2}{2\left(1 + u^2\right)} - \tfrac{1}{2}\arctan u - \frac{u}{2\left(1 + u^2\right)} + C$$

$$= \frac{x^2 + 3x + 2}{2\left(x^2 + 4x + 5\right)} - \tfrac{1}{2}\arctan(x+2) + C.$$

7. First, $3 - 2x - x^2 = 4 - (x+1)^2$, so we let $x = -1 + 2\sin u$: $x + 1 = 2\sin u$, $dx = 2\cos u\, du$. Then $4 - (x+1)^2 = 4 - 4\sin^2 u = 4\cos^2 u$, so

$$\int x\sqrt{3 - 2x - x^2}\, dx = \int (-1 + 2\sin u)(2\cos u)(2\cos u)\, du$$

$$= \int \left(-4\cos^2 u + 8\cos^2 u \sin u\right)\, du$$

$$= \int \left(-2 - 2\cos 2u + 8\cos^2 u \sin u\right)\, du$$

$$= -2u - 2\sin u \,\cos u - \tfrac{8}{3}\cos^3 u + C$$

$$= -2\arcsin\left(\frac{x+1}{2}\right) - \tfrac{1}{2}(x+1)\sqrt{3 - 2x - x^2} - \tfrac{1}{3}\left(3 - 2x - x^2\right)^{3/2} + C.$$

10. Because $4x^2 + 4x - 3 = (2x+1)^2 - 4$, we let $x = -\frac{1}{2} + \sec u$: $dx = \sec u \tan u\,du$, $2x + 1 = 2\sec u$, and $(2x+1)^2 - 4 = 4\tan^2 u$.

$$\int \sqrt{4x^2 + 4x - 3}\,dx = \int 2\sec u \tan^2 u\,du = 2\int (\sec^3 u - \sec u)\,du$$
$$= \sec u \tan u - \ln|\sec u + \tan u| + C$$
$$= \tfrac{1}{4}(2x+1)(4x^2+4x-3)^{1/2} - \ln\left|2x+1+\sqrt{4x^2+4x-3}\,\right| + C$$

(the constant $\ln 2$ has been "absorbed" by C).

13. $3 + 2x - x^2 = 4 - (x-1)^2 = 4 - 4\sin^2 u$ if we let $x = 1 + 2\sin u$. Then $dx = 2\cos u\,du$, and therefore

$$\int \frac{1}{3+2x-x^2}\,dx = \int \frac{2\cos u}{4\cos^2 u}\,du = \tfrac{1}{2}\int \sec u\,du$$
$$= \tfrac{1}{2}\ln|\sec u + \tan u| + C = \tfrac{1}{2}\ln\left|\frac{x+1}{\sqrt{3+2x-x^2}}\right| + C.$$

16. Since $4x^2 + 4x - 15 = (2x+1)^2 - 16$, we let $2x + 1 = 4\sec z$. Then $x = \tfrac{1}{2}(-1 + 4\sec z)$ and $dx = 2\sec z \tan z\,dz$.

$$\int \frac{2x-1}{4x^2+4x-15}\,dx = \int \frac{4\sec z - 2}{16\tan^2 z}(2\sec z \tan z)\,dz$$
$$= \tfrac{1}{4}\int \frac{(2\sec z - 1)\sec z}{\tan z}\,dz$$
$$= \tfrac{1}{2}\int \frac{\sec^2 z}{\tan z}\,dz - \tfrac{1}{4}\int \csc z\,dz$$
$$= \tfrac{1}{2}\ln|\tan z| - \tfrac{1}{4}\ln|\csc z - \cot z| + C_1.$$

With the aid of the reference triangle to the right, we can "translate" the expression above into a function of the original variable x. Thus we find the antiderivative to be

$$\tfrac{1}{2}\ln\frac{\sqrt{4x^2+4x-15}}{4} - \tfrac{1}{4}\ln\frac{2x+1-4}{\sqrt{4x^2+4x-15}} + C_1,$$

which can be simplified to

$$\tfrac{1}{8}\ln|2x-3| - \tfrac{3}{8}\ln|2x+5| + C.$$

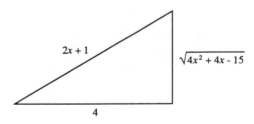

Triangle with legs labeled $2x+1$ and 4, and hypotenuse $\sqrt{4x^2 + 4x - 15}$.

19. First, $9 + 16x - 4x^2 = 25 - (2x-4)^2$. Then use the trigonometric substitution $x = 2 + \tfrac{5}{2}\sin u$.

$$\int (7-2x)\sqrt{9+16x-4x^2} = \int \left(\tfrac{75}{2}\cos^2 u - \tfrac{125}{2}\cos^2 u \sin u\right)du$$
$$= \tfrac{25}{12}(9u + 9\sin u \cos u + 10\cos^3 u) + C.$$

22. Let $x = \tan u$. Then

$$\int \frac{x-1}{(x^2+1)^2}\,dx = \int \frac{\tan u - 1}{\sec^4 u}\sec^2 u\,du = \int \left(\frac{\tan u}{\sec^2 u} - \frac{1}{\sec^2 u}\right)du = \int (\cos u \sin u - \cos^2 u)\,du$$
$$= \tfrac{1}{2}\sin^2 u - \tfrac{1}{2}u - \tfrac{1}{2}\sin u \cos u + C$$
$$= \frac{x^2}{2(x^2+1)} - \tfrac{1}{2}\tan^{-1} x - \frac{x}{2(x^2+1)} + C$$
$$= \frac{x(x-1)}{2(x^2+1)} - \tfrac{1}{2}\tan^{-1} x + C.$$

25.   We note that $x^2 + x + 1 = \frac{3}{4}\left(\frac{4}{3}\left(x + \frac{1}{2}\right)^2 + 1\right)$. This suggests that we attempt to obtain

$$\tfrac{4}{3}\left(x + \tfrac{1}{2}\right)^2 = \tan^2 u,$$

which we accomplish by using the substitution

$$x = \tfrac{1}{2}\left(-1 + \sqrt{3}\tan u\right).$$

This leads to

$$\int \frac{3x - 1}{x^2 + x + 1}\, dx = \int \left(3\tan u - \tfrac{5}{3}\sqrt{3}\right)\, du,$$

and only the final answer is complicated.

28.   Because $x - x^2 = \frac{1}{4}\left(1 - (2x - 1)^2\right)$, we want $2x - 1 = \sin u$: Let $x = \frac{1}{2}(1 + \sin u)$. This yields

$$\int (x - x^2)^{2/3}\, dx = \tfrac{1}{16}\int \cos^4 u\, du.$$

Then we apply Formula 34 of the endpapers to obtain

$$\int (x - x^2)^{3/2}\, dx = \tfrac{1}{128}\left(3u + 4\sin u\cos u + \sin u\cos^3 u - \sin^3 u\cos u\right) + C$$

$$= \tfrac{3}{128}\sin^{-1}(2x - 1) - \tfrac{1}{64}\left(16x^3 - 24x^2 + 2x + 3\right)\sqrt{x - x^2} + C.$$

31. $\dfrac{2x^2 + 3}{x^4 - 2x^2 + 1} = \dfrac{A}{x - 1} + \dfrac{B}{(x - 1)^2} + \dfrac{C}{x + 1} + \dfrac{D}{(x + 1)^2}$ leads to $A = -\tfrac{1}{4}$, $B = \tfrac{5}{4}$, $C = \tfrac{1}{4}$, and $D = \tfrac{5}{4}$, and thus:

$$\int \frac{2x^2 + 3}{x^4 - 2x^2 + 1} = \tfrac{1}{4}\ln\left|\frac{x + 1}{x - 1}\right| - \frac{5x}{2(x^2 - 1)} + C.$$

34. $\dfrac{x^3 - 2x}{(x^2 + 2x + 2)^2} = \dfrac{Ax + B}{x^2 + 2x + 2} + \dfrac{Cx + D}{(x^2 + 2x + 2)^2}$ yields $A = 1$, $B = -2$, $C = 0$, and $D = 4$. This leads to two integrals:

$$\int \frac{x - 2}{(x + 1)^2 + 1}\, dx \quad\text{and}\quad \int \frac{4}{\left((x + 1)^2 + 1\right)^2}\, dx.$$

Let $u = x + 1$. The first integral is then easy to evaluate:

$$\int \frac{u - 3}{u^2 + 1}\, du = \tfrac{1}{2}\int \frac{2u}{u^2 + 1}\, du - 3\int \frac{du}{u^2 + 1}$$

$$= \tfrac{1}{2}\ln\left(u^2 + 1\right) - 3\arctan u + C$$

$$= \tfrac{1}{2}\ln\left(x^2 + 2x + 2\right) - 3\arctan(x + 1) + C.$$

The second integral becomes

$$\int \frac{4}{(u^2 + 1)^2}\, du,$$

and the substitution $u = \tan z$ works well. The answer to the original problem is

$$\int \frac{x^3 - 2x}{x^2 + 2x + 2}\, dx = \tfrac{1}{2}\ln\left(x^2 + 2x + 2\right) - \arctan(x + 1) + \frac{2(x + 1)}{x^2 + 2x + 2} + C.$$

37.   First, $y = \frac{19}{4} + \sqrt{R^2 - (x + 1)^2}$ where $R = \sqrt{377/16}$. It follows that $\dfrac{dy}{dx} = -\dfrac{x + 1}{\sqrt{R^2 - (1 + x)^2}}$ and hence the length of the road is given by

$$S = \int_0^3 R\left(R^2 - (1 + x)^2\right)^{-1/2}\, dx$$

$$= R\left[\sin^{-1}\left(\frac{1 + x}{R}\right)\right]_0^3 = -\tfrac{1}{4}\sqrt{377}\arctan\tfrac{20}{21} \approx 3.6940487.$$

Thus the length of the road is just over 3.69 miles.

40. $\dfrac{1}{x^3+8} = \dfrac{A}{x+2} + \dfrac{Bx+C}{x^2-2x+4}$ leads to $A = \frac{1}{12}$, $B = -\frac{1}{12}$, and $C = \frac{1}{3}$. Thus

$$\int \frac{1}{x^3+8}\,dx = \frac{1}{12}\int\left(\frac{1}{x+2} + \frac{4-x}{x^2-2x+4}\right)dx$$

$$= \frac{1}{12}\int\left(\frac{1}{x+2} - \frac{1}{2}\left(\frac{2x-2}{x^2-2x+4}\right) + \frac{3}{x^2-2x+4}\right)dx.$$

Note that $x^2 - 2x + 4 = (x-1)^2 + 3$, and with the aid of Formula 17 of the endpapers we obtain the final answer; after simplification, it is

$$\int \frac{1}{x^3+8}\,dx = \frac{1}{24}\ln\left|\frac{x^2+4x+4}{x^2-2x+4}\right| + \frac{1}{12}\sqrt{3}\,\arctan\left(\frac{1}{3}\sqrt{3}\,(x-1)\right) + C.$$

43. $x^4 + x^2 + 1 = (x^2+x+1)(x^2-x+1)$. (It follows that $10^{12} + 10^6 + 1 = 1,000,001,000,001$ is composite.) And

$$\frac{2x^3+3x}{x^4+x^2+1} = \frac{Ax+B}{x^2+x+1} + \frac{Cx+D}{x^2-x+1}$$

yields the values $A = 1$, $B = -\frac{1}{2}$, $C = 1$, and $D = \frac{1}{2}$. So

$$\int \frac{2x^3+3x}{x^4+x^2+1}\,dx = \frac{1}{2}\int\left(\frac{2x-1}{x^2+x+1} + \frac{2x+1}{x^2-x+1}\right)dx$$

$$= \frac{1}{2}\int\left(\frac{2x+1}{x^2+x+1} - \frac{2}{x^2+x+1} + \frac{2x-1}{x^2-x+1} + \frac{2}{x^2-x+1}\right)dx$$

$$= \frac{1}{2}\left(\ln(x^2+x+1) + \ln(x^2-x+1)\right) + \int\left(\frac{1}{\left(x-\frac{1}{2}\right)^2+\frac{3}{4}} - \frac{1}{\left(x+\frac{1}{2}\right)^2+\frac{3}{4}}\right)dx$$

$$= \frac{1}{2}\ln(x^4+x^2+1) + \frac{2}{3}\sqrt{3}\int\left(\frac{\frac{2}{3}\sqrt{3}}{\left(\frac{2x-1}{\sqrt{3}}\right)^2+1} - \frac{\frac{2}{3}\sqrt{3}}{\left(\frac{2x+1}{\sqrt{3}}\right)^2+1}\right)dx$$

$$= \frac{1}{2}\ln(x^4+x^2+1) + \frac{2}{3}\sqrt{3}\left(\arctan\left(\frac{2x-1}{\sqrt{3}}\right) - \arctan\left(\frac{2x+1}{\sqrt{3}}\right)\right) + C$$

$$= \frac{1}{2}\ln(x^4+x^2+1) + \frac{2\sqrt{3}}{3}\arctan\left(\frac{\sqrt{3}}{3}(2x^2+1)\right) + C.$$

Section 9.8

1. $\displaystyle\int_4^\infty x^{-3/2}\,dx = \lim_{t\to\infty}\left[\frac{2}{x^{1/2}}\right]_t^4 = 1.$

4. $\displaystyle\int_0^8 x^{-2/3}\,dx = \lim_{t\to 0}\left[3x^{1/3}\right]_t^8 = 6.$

7. $\displaystyle\int_5^\infty (x-1)^{-3/2}\,dx = \lim_{t\to\infty}\left[-\frac{2}{\sqrt{x-1}}\right]_5^t = 1.$

10. $\displaystyle\int_0^3 \frac{1}{(x-3)^2}\,dx = \lim_{t\to 3^-}\left[-\frac{1}{x-3}\right]_0^t$ diverges.

13. Integrate from -1 to 0, from 0 to 8. If both integrals converge add the results.

16. $\displaystyle\int_{-\infty}^{\infty}\frac{x}{(x^2+4)^{3/2}}\,dx=\int_{-\infty}^{0}\frac{x}{(x^2+4)^{3/2}}\,dx+\int_{0}^{\infty}\frac{x}{(x^2+4)^{3/2}}\,dx$

$$=\lim_{s\to-\infty}\left[-\frac{1}{\sqrt{x^2+4}}\right]_{s}^{0}+\lim_{t\to\infty}\left[-\frac{1}{\sqrt{x^2+4}}\right]_{0}^{t}=0.$$

22. $\displaystyle\int_{0}^{\infty}\sin^2 x\,dx$ diverges because $\displaystyle\int_{0}^{\pi}\sin^2 x\,dx=\frac{\pi}{2}$.

28. $\displaystyle\int_{0}^{\infty}\frac{1+x}{1+x^2}\,dx$ dominates $\displaystyle\int_{0}^{\infty}\frac{x}{1+x^2}\,dx$, which diverges.

34. The force exerted on m by dy is $\dfrac{Gm\rho}{a^2+y^2}\,dy$. The horizontal component of that force is

$$dF=\frac{Gm\rho}{r^2}\,(\cos\theta)\,dy.$$

So the total horizontal force is

$$F=\int_{-\infty}^{\infty}\frac{Gm\rho}{r^2}\,(\cos\theta)\,dy.$$

Now $r^2=a^2+y^2$ and $\cos\theta=\dfrac{a}{r}$. So $\cos\theta=\dfrac{a}{\sqrt{a^2+y^2}}$: $\;F=2\displaystyle\int_{0}^{\infty}\frac{Gm\rho a}{(a^2+y^2)^{3/2}}\,dy.$

With the aid of Formula 52 of the endpapers of the text, we find that

$$F=2Gm\rho a\left[\frac{y}{a^2\sqrt{a^2+y^2}}\right]_{0}^{\infty}=\frac{2Gm\rho}{a}\left(\lim_{y\to\infty}\frac{y}{\sqrt{a^2+y^2}}\right)=\frac{2Gm\rho}{a}.$$

37. (a)
$$\begin{bmatrix}u=x^{k-1} & dv=xe^{-x^2}\,dx\\[4pt] du=(k-1)\,x^{k-2}\,dx & v=-\tfrac{1}{2}e^{-x^2}\end{bmatrix}$$

$$\int_{0}^{\infty}x^k\,e^{-x^2}\,dx=\left[-\frac{x^{k-1}}{2}\,e^{-x^2}\right]_{0}^{\infty}+\frac{k-1}{2}\int_{0}^{\infty}x^{k-2}\,e^{-x^2}\,dx=\frac{k-1}{2}\int_{0}^{\infty}x^{k-2}\,e^{-x^2}\,dx.$$

(b) If $n=1$, then

$$\int_{0}^{\infty}x^{n-1}\,e^{-x^2}\,dx=\int_{0}^{\infty}e^{-x^2}\,dx=\frac{1}{2}\sqrt{\pi}\quad\text{(by Problem 35)}$$

$$=\frac{1}{2}\Gamma\left(\frac{1}{2}\right)\quad\text{(also by Problem 35).}$$

If $n=2$, then

$$\int_{0}^{\infty}x^{n-1}\,e^{-x^2}\,dx=\int_{0}^{\infty}xe^{-x^2}\,dx=\left[-\frac{1}{2}e^{-x^2}\right]_{0}^{\infty}=\frac{1}{2}=\frac{1}{2}\Gamma(1)\text{ because }\Gamma(1)=0!=1.$$

If $n=3$, then

$$\int_{0}^{\infty}x^{n-1}\,e^{-x^2}\,dx=\frac{n-2}{2}\int_{0}^{\infty}x^{n-3}\,e^{-x^2}\,dx\text{ (by Part (a))}$$

$$=\frac{1}{2}\int_{0}^{\infty}e^{-x^2}\,dx=\frac{1}{2}\cdot\frac{1}{2}\sqrt{\pi}$$

$$=\frac{1}{2}\Gamma\left(\frac{3}{2}\right)\text{ because }\Gamma\left(\frac{3}{2}\right)=\frac{1}{2}\Gamma\left(\frac{1}{2}\right)=\frac{1}{2}\sqrt{\pi}.$$

Assume that for some integer $k\geq 3$,

$$\int_{0}^{\infty}x^{n-1}e^{-x^2}\,dx=\frac{1}{2}\Gamma\left(\frac{n}{2}\right)\text{ for all }n,\;1\leq n\leq k.$$

Then

$$\int_0^\infty x^k\, e^{-x^2}\, dx = \frac{k-1}{2} \int_0^\infty x^{k-2} e^{-x^2}\, dx = \frac{k-1}{2} \cdot \frac{1}{2} \Gamma\left(\frac{k-1}{2}\right) = \frac{1}{2}\Gamma\left(\frac{k+2}{2}\right)$$

(because $\Gamma(x+1) = x\,\Gamma(x)$). Therefore, by induction, $\displaystyle\int_0^\infty x^{n-1} e^{-x^2}\, dx = \frac{1}{2}\Gamma\left(\frac{n}{2}\right)$ for all $n \geq 1$.

Chapter 9 Miscellaneous

4. $\displaystyle\int \frac{\csc x \cot x}{1 + \csc^2 x}\, dx = -\tan^{-1}(\csc x) + C$

7. Integration by parts: Let $u = x$, $dv = \tan^2 x\, dx = (\sec^2 x - 1)\, dx$.

10. Let $x = 2\tan u$:

$$\int \frac{1}{\sqrt{x^2+4}}\, dx = \int \sec u\, du = \ln|\sec u + \tan u| + C$$
$$= \ln\left|\tfrac{1}{2}\sqrt{x^2+4} + \tfrac{1}{2}x\right| + C_1 = \ln\left|x + \sqrt{x^2+4}\right| + C.$$

13. Write $x^2 - x + 1 = \left(x - \tfrac{1}{2}\right)^2 + \tfrac{3}{4}$, then apply Formula 17 from the endpapers.

16. $\displaystyle\int \frac{x^4+1}{x^2+2}\, dx = \int \left(x^2 - 2 + \frac{5}{x^2+2}\right)\, dx = \tfrac{1}{3}x^3 - 2x + \tfrac{5}{2}\sqrt{2}\tan^{-1}\left(\tfrac{1}{2}x\sqrt{2}\right) + C.$

19. Use the substitution $u = \sin x$.

22. Let $x^2 = \sin u$:

$$\int \frac{x^7}{\sqrt{1-x^4}}\, dx = \int \tfrac{1}{2}\sin^3 u\, du = \tfrac{1}{6}(\cos u)(\cos^2 u - 3) + C = -\tfrac{1}{6}\left(x^4 + 2\right)\sqrt{1-x^4} + C.$$

25. Let $x = 3\tan u$.

28. $\displaystyle\int \frac{4x-2}{x^3-x}\, dx = \int \left(\frac{2}{x} - \frac{3}{x+1} + \frac{1}{x-1}\right)\, dx$

$$= 2\ln|x| - 3\ln|x+1| + \ln|x-1| + C = \ln\left|\frac{x^2(x-1)}{(x+1)^3}\right| + C.$$

31. Let $x = -1 + \tan u$: $\displaystyle\int \frac{x}{(x^2+2x+2)^2}\, dx = \int \frac{-1+\tan u}{\sec^2 u}\, du = \int \left(-\cos^2 u + \sin u \cos u\right)\, du$, and the rest is routine.

34. $\displaystyle\int \frac{\sec x}{\tan x}\, dx = \int \csc x\, dx = \ln|\csc x - \cot x| + C.$

37. *Suggestion:* Develop a reduction formula for $\displaystyle\int x\,(\ln x)^n\, dx$ by parts; take $u = (\ln x)^n$ and $dv = x\, dx$. Then apply your formula iteratively to evaluate the given antiderivative. You should find that

$$\int x\,(\ln x)^n\, dx = \tfrac{1}{2}x^2 (\ln x)^n - \frac{n}{2}\int x\,(\ln x)^{n-1}\, dx.$$

40. Note that $4x - x^2 = 4 - (x-2)^2$; let $x - 2 = 2\sin u$. Then

$$\int \frac{x}{\sqrt{4x - x^2}}\, dx = \int 2(1 + \sin u)\, du = 2(u - \cos u) + C = 2\arcsin\frac{x-2}{2} - \sqrt{4x - x^2} + C.$$

43. Use the method of partial fraction decomposition.

46. Here is a quick way to obtain the partial fraction decomposition: $\displaystyle\frac{x^2+2x+2}{(x+1)^3} = \frac{(x+1)^2 + 1}{(x+1)^3}.$

$$\int \frac{x^2+2x+2}{(x+1)^3}\, dx = \int \left(\frac{1}{x+1} + \frac{1}{(x+1)^3}\right)\, dx = \ln|x+1| - \frac{1}{2(x+1)^2} + C.$$

49. The partial fraction decomposition of the integrand has the form

$$\frac{A}{x-1} + \frac{Bx+C}{x^2+x+1} + \frac{D}{(x-1)^2} + \frac{Ex+F}{(x^2+x+1)^2}.$$

The simultaneous equations are

$$
\begin{aligned}
A + B &= 3 \\
A - B + C + D &= -1 \\
A \quad\ - C + 2D + E &= 2 \\
-A - B \quad\ + 3D - 2E + F &= -12 \\
-A + B - C + 2D + E - 2F &= -2 \\
-A \quad\ + C + D \quad\ + F &= 1
\end{aligned}
$$

Their solution: $A = 1$, $B = 2$, $C = 1$, $D = -1$, $E = 4$, and $F = 2$. None of the antiderivatives is difficult.

$$\int \frac{3x^5 - x^4 + 2x^3 - 12x^2 - 2x + 1}{(x^3 - 1)^2}\, dx = \ln|x-1| + \ln(x^2+x+1) + \frac{1}{x-1} - \frac{2}{x^2+x+1} + C.$$

52. Let $x = u^3$: $\displaystyle\int \frac{(1+x^{2/3})^{3/2}}{x^{1/3}}\, dx = 3\int u\,(1+u^2)^{3/2}\, du = \tfrac{3}{5}(1+u^2)^{5/2} + C = \tfrac{3}{5}(1+x^{2/3})^{5/2} + C.$

55. $\tan^3 z = (\tan z)(\sec^2 z - 1) = (\tan z)(\sec^2 z) - \tan z.$

58. Note that $\dfrac{\cos^3 x}{\sqrt{\sin x}} = \dfrac{(1-\sin^2 x)(\cos x)}{\sqrt{\sin x}} = (\sin x)^{-1/2}\cos x - (\sin x)^{3/2}\cos x.$ So

$$\int \frac{\cos^3 x}{\sqrt{\sin x}}\, dx = \tfrac{2}{5}(\sin x)^{1/2}(5 - \sin^2 x) + C.$$

61. Use integration by parts with $u = \arcsin x$ and $dv = \dfrac{1}{x^2}\, dx$. Then apply Formula 60 of the endpapers, or use the trigonometric substitution $x = \sin u$.

64. Because $2x - x^2 = 1 - (x-1)^2$, let $x = 1 + \sin u$:

$$\int x\sqrt{2x - x^2}\, dx = \int (1 + \sin u)(\cos^2 u)\, du = \int \left(\frac{1 + \cos 2u}{2} + \cos^2 u \sin u\right) du$$

$$= \tfrac{1}{2}u + \tfrac{1}{2}\sin u \cos u - \tfrac{1}{3}\cos^3 u + C$$

$$= \tfrac{1}{2}\sin^{-1}(x-1) + \tfrac{1}{2}(x-1)\sqrt{2x-x^2} - \tfrac{1}{3}(2x - x^2)^{3/2} + C$$

$$= \tfrac{1}{2}\sin^{-1}(x-1) + \tfrac{1}{6}\sqrt{2x-x^2}(2x^2 - x - 3) + C.$$

70. Let $u = \tan x$: $\displaystyle\int \frac{\sec^2 x}{\tan^2 x + 2\tan x + 2}\, dx = \int \frac{1}{u^2 + 2u + 2}\, du = \int \frac{1}{1 + (u+1)^2}\, du$

$$= \tan^{-1}(u+1) + C = \tan^{-1}(1 + \tan x) + C.$$

73. Let $u = x^3 - 1$; the rest is routine.

76. Let $x = \tan^3 z$. $\displaystyle\int \frac{1}{x^{2/3}(1 + x^{2/3})}\, dx = \int \frac{3\tan^2 z \sec^2 z}{\sec^2 z \tan^2 z}\, dz = 3z + C = 3\tan^{-1}(x^{1/3}) + C.$

79. Multiply numerator and denominator of the integrand by $\sqrt{1 - \sin t}$.

82. Let $u = e^x$: $\displaystyle\int e^x \sin^{-1}(e^x)\, dx = \int \sin^{-1} u\, du.$ Now do an integration by parts with $p = \sin^{-1} u$, $dq = du$:

$$\int e^x \sin^{-1}(e^x)\, dx = u\sin^{-1} u - \int u(1 - u^2)^{-1/2}\, du$$

$$= u\sin^{-1} u + \sqrt{1 - u^2} + C = e^x \sin^{-1}(e^x) + \sqrt{1 - e^{2x}} + C.$$

85. The partial fraction decomposition of the integrand is $\dfrac{x}{x^2+1} - \dfrac{x}{(x^2+1)^2}$, and integration of these terms presents no difficulties.

88. $\left[\begin{array}{ll} u = \ln x & dv = x^{3/2}\,dx \\ du = \dfrac{1}{x}\,dx & v = \frac{2}{5}x^{5/2} \end{array}\right]$ $\qquad$ $\displaystyle\int x^{3/2}\ln x\,dx = \frac{2}{25}x^{5/2}(-2 + 5\ln x) + C.$

91. Use integration by parts. A good choice is $u = x$, $dv = e^x \sin x\,dx$. One must then antidifferentiate both $e^x \sin x$ and $e^x \cos x$, but here Formulas 67 and 68 of the endpapers may be used, or integration by parts will suffice for each.

94. $\left[\begin{array}{ll} u = \ln(1 + \sqrt{x}) & dv = dx \\ du = \dfrac{1}{2\sqrt{x}(1+\sqrt{x})}\,dx & v = x - 1 \end{array}\right]$ $\displaystyle\int \ln(1+\sqrt{x})\,dx = (x-1)\ln(1+\sqrt{x}) - \tfrac{1}{2}x + \sqrt{x} + C.$

97. Let $u = x - 1$: $\displaystyle\int \frac{x^4}{(x-1)^2}\,dx = \int \frac{(u+1)^4}{u^2}\,du = \int\left(u^2 + 4u + 6 + \frac{4}{u} + \frac{1}{u^2}\right)du.$

100. Let $u = x^2$: $\displaystyle\int x\sqrt{\frac{1-x^2}{1+x^2}}\,dx = \frac{1}{2}\int\sqrt{\frac{1-u}{1+u}}\,du.$

$\quad$ Now let $v^2 = \dfrac{1-u}{1+u}$; $u = \dfrac{1-v^2}{1+v^2}$ and $du = -\dfrac{4v}{(1+v^2)^2}\,dv.$

$$\int x\sqrt{\frac{1-x^2}{1+x^2}}\,dx = -\frac{1}{2}\int \frac{4v^2}{(1+v^2)^2}\,dv.$$

$\quad$ Finally let $v = \tan z$.

$$\int x\sqrt{\frac{1-x^2}{1+x^2}}\,dx = -2\int \sin^2 z\,dz = \sin z\,\cos z - z + C = \frac{1}{2}\sqrt{1-x^4} - \tan^{-1}\frac{1-x^2}{\sqrt{1+x^2}} + C.$$

103. $A_t = \displaystyle\int_0^t 2\pi e^{-x}\sqrt{1 + e^{-2x}}\,dx.$ Let $u = e^{-x}$. Then

$$A_t = \int_1^{e^{-t}} 2\pi u\sqrt{1+u^2}\left(-\frac{1}{u}\right)du = \int_p^1 2\pi\sqrt{1+u^2}\,du \quad(\text{where } p = e^{-t}).$$

$\quad$ Let $u = \tan z$. Then

$$A_t = \int_{u=p}^{u=1} 2\pi\sqrt{1+\tan^2 z}\,\sec^2 z\,dz$$

$$= \pi\Big[\sec z\,\tan z + \ln|\sec z + \tan z|\Big]_{u=p}^{u=1}$$

$$= \pi\left(\sqrt{2} + \ln\left(1+\sqrt{2}\right) - e^{-t}\sqrt{1+e^{-2t}} - \ln\left|e^{-t} + \sqrt{1+e^{-2t}}\right|\right).$$

$\quad \displaystyle\lim_{t\to\infty} A_t = \pi\left(\sqrt{2} + \ln\left(1+\sqrt{2}\right)\right) \approx 7.2118.$

106. (a): Use integration by parts with $u = (\ln x)^n$ and $dv = x^m\,dx$.

$\quad$ (b): Application of the formula and simplification of the result yields $\dfrac{17e^4 + 3}{128} \approx 7.2747543.$

109. The area is $A = 2\displaystyle\int_0^2 x^{5/2}(2-x)^{1/2}\,dx.$ Use the suggested substitution $x = 2\sin^2\theta$. This results

$\quad$ in $A = 64\displaystyle\int_0^{\pi/2} \sin^6\theta\,\cos^2\theta\,d\theta.$ Then the formula of Problem 108 gives the answer $\frac{5}{4}\pi.$

112. $\sqrt{1 + (dy/dx)^2} = \sqrt{1 + \sqrt{x}}$; the curve's length is $L = \displaystyle\int_0^1 \sqrt{1 + \sqrt{x}}\ dx$. Let $x = \tan^4 u$.

$$L = \int_0^{\pi/4} 4\sqrt{1 + \tan^2 u}\ \tan^3 u \sec^2 u\ du = 4\int_0^{\pi/4} \sec^3 u \tan^3 u\ du$$

$$= 4\int_0^{\pi/4} \left(\sec^3 u\right)\left(\sec^2 u - 1\right)\tan u\ du = 4\int_0^{\pi/4} \left(\sec^4 u - \sec^2 u\right)\left(\sec u \tan u\right)du$$

$$= \left[\tfrac{4}{5}\sec^5 u - \tfrac{4}{3}\sec^3 u\right]_0^{\pi/4} = \tfrac{4}{5}\left(4\sqrt{2} - 1\right) - \tfrac{4}{3}\left(2\sqrt{2} - 1\right) = \tfrac{8}{15}\left(1 + \sqrt{2}\right) \approx 1.28758.$$

115. Let $u = e^x$. $\displaystyle\int \frac{1}{1 + e^x + e^{-x}}\ dx = \int \frac{1}{u^2 + u + 1}\ du.$

118. With the recommended substitution, the numerator is $\frac{1}{2}\ du$, so

$$\int \frac{1 + 2x^2}{x^5\left(1 + x^2\right)^3}\ dx = \tfrac{1}{2}\int u^{-3}\ du = -\tfrac{1}{4}u^{-2} + C = -\frac{1}{4\left(x^4 + x^2\right)^2} + C.$$

Note: A partial fraction decomposition results in

$$\int \frac{1 + 2x^2}{x^5\left(1 + x^2\right)^3}\ dx = \frac{1}{2x^2} - \frac{1}{4x^4} - \frac{1}{2\left(x^2 + 1\right)} - \frac{1}{4\left(x^2 + 1\right)^2} + C.$$

121. Let $3x - 2 = u^2$: $\displaystyle\int x^3\sqrt{3x - 2}\ dx = \int \left(\frac{u^2 + 2}{3}\right)^3 \cdot u \cdot \tfrac{2}{3}u\ du$

124. Let $x - 1 = u^2$:

$$\int x^2\left(x - 1\right)^{3/2}\ dx = \int \left(2u^8 + 4u^6 + 2u^4\right)du = \tfrac{2}{9}\left(x - 1\right)^{9/2} + \tfrac{4}{7}\left(x - 1\right)^{7/2} + \tfrac{2}{5}\left(x - 1\right)^{5/2} + C$$

130. $\displaystyle\int \sqrt{1 + \sqrt{x}}\ dx = \tfrac{4}{15}\left(3\sqrt{x} - 2\right)\left(1 + \sqrt{x}\right)^{3/2} + C.$

133. Area: $A = 2\displaystyle\int_0^1 x\sqrt{1 - x}\ dx$. For variety, let $x = \sin^2 \theta$.

$$A = 2\int_0^{\pi/2} \left(\sin^2 \theta\right)\left(\cos \theta\right)\left(2\sin \theta \cos \theta\right)d\theta$$

$$= 4\int_0^{\pi/2} \left(\cos^2 \theta - \cos^4 \theta\right)\sin \theta\ d\theta$$

$$= 4\left[\tfrac{1}{5}\cos^5 \theta - \tfrac{1}{3}\cos^3 \theta\right]_0^{\pi/2} = \tfrac{8}{15}.$$

136. $\displaystyle\int \frac{1}{5 + 4\cos \theta}\ d\theta = \int \frac{2}{u^2 + 9}\ du = \tfrac{2}{3}\arctan\left(\frac{u}{3}\right) + C = \tfrac{2}{3}\arctan\left(\tfrac{1}{3}\tan\frac{\theta}{2}\right) + C.$

139. $\displaystyle\int \frac{1}{\sin \theta + \cos \theta}\ d\theta = \int \frac{2}{1 + 2u - u^2}\ du = \frac{\sqrt{2}}{2}\ln\left|\frac{u + \sqrt{2} - 1}{u - \sqrt{2} - 1}\right| + C.$

142. $\displaystyle\int \frac{\sin \theta - \cos \theta}{\sin \theta + \cos \theta}\ d\theta = -2\int \frac{u^2 + 2u - 1}{\left(u^2 + 1\right)\left(u^2 - 2u - 1\right)}\ du$

$$= \ln\left|\frac{u^2 + 1}{u + \sqrt{2} - 1}\right| - \ln\left|u - \sqrt{2} - 1\right| + C = -\ln\left|\sin \theta + \cos \theta\right| + C.$$

Chapter 10: Polar Coordinates and Conic Sections

Section 10.1

1. The given line has slope $-\frac{1}{2}$.

4. By implicit differentiation, $\dfrac{dy}{dx} = \dfrac{1}{2y}$; in particular, the slope of the tangent at $(6, -3)$ is $-\frac{1}{6}$. So the tangent line has equation $y + 3 = -\frac{1}{6}(x - 6)$; that is, $x + 6y + 12 = 0$.

7. Write the equation in the form $x^2 + 2x + 1 + y^2 = 5$.

10. Write the equation in the form $x^2 + 8x + 16 + y^2 - 6y + 9 = 25$ to conclude that the center is at $(-4, 3)$ and the radius is 5.

13. Complete the squares as follows: $2x^2 + 2y^2 - 2x + 6y = 13$;
$$x^2 + y^2 - x + 3y = \tfrac{13}{2};$$
$$x^2 - x + \tfrac{1}{4} + y^2 + 3y + \tfrac{9}{4} = \tfrac{36}{4};$$
$$\left(x - \tfrac{1}{2}\right)^2 + \left(y + \tfrac{3}{2}\right)^2 = 3^2.$$
Thus the center is at $\left(\tfrac{1}{2}, -\tfrac{3}{2}\right)$ and the radius is 3.

16. Center: $\left(\tfrac{2}{3}, \tfrac{3}{2}\right)$; radius: 12.

19. Complete the squares as follows: $x^2 + y^2 - 6x - 10y + 84 = 0$;
$$x^2 - 6x + 9 + y^2 - 10y + 25 + 50 = 0;$$
$$(x - 3)^2 + (y - 5)^2 = -50.$$
It follows that there are no points on the graph.

22. The line has slope 1, so the radius to the point of tangency has slope -1. The radius is a segment of the line with equation $y + 2 = -(x - 2)$; that is, $x + y = 0$. The point of tangency is the simultaneous solution of the equations of the two lines: $x = -2$, $y = 2$—therefore they meet at $(-2, 2)$. The length of the radius is the distance from $(2, -2)$ to $(-2, 2)$, which is $4\sqrt{2}$. Therefore the equation of the circle is $(x - 2)^2 + (y + 2)^2 = 32$.

25. The squares of the distances are equal, so $P(x, y)$ satisfies $(x - 3)^2 + (y - x)^2 = (x - 7)^2 + (y - 4)^2$. Expand and simplify to obtain the answer.

28. The point $P(x, y)$ satisfies the equation $x + 3 = \sqrt{(x - 3)^2 + y^2}$. Expand and simplify to obtain $y^2 = 12x$: The locus is a parabola, opening to the right, with its vertex at the origin, and symmetric about the x-axis.

31. If $P(a, b)$ is the point of tangency, then $b = a^2$, and the slope of the tangent line can be measured in two different ways. They are equal, and we thereby obtain
$$\frac{1 - a^2}{2 - a} = 2a.$$
We find two solutions: $a = 2 - \sqrt{3}$ and $a = 2 + \sqrt{3}$. But $b = a^2$, so $b = 7 - 4\sqrt{3}$ or $b = 7 - 4\sqrt{3}$. Thus we obtain the two answers given in the text.

34. The second condition means that all such lines have slope 3. If such a line is tangent to the graph of $y = x^3$ at the point (a, a^3), its slope must also be $3a^2 = 3$. Thus $a = 1$ or $a = -1$. Thus there are two such lines:
$$y - 1 = 3(x - 1) \quad \text{(through } (1, 1) \text{ with slope 3)} \quad \text{and}$$
$$y + 1 = 3(x + 1) \quad \text{(through } (-1, -1) \text{ with slope 3)}.$$

4.   Answer: $r \sin \theta = 6$; a better form is $r = 6 \csc \theta, 0 < \theta < \pi$.

10. Answer: $r = \dfrac{4}{\sin \theta + \cos \theta}$. It would be best to restrict the values of θ to the range $-\pi/4 < \theta < 3\pi/4$.

13. Begin by multiplying each side of the equation by r.

16. First, $r^2 = 2r + r \sin \theta$, so $x^2 + y^2 = 2\sqrt{x^2 + y^2} + y$. Hence

$$\left(x^2 + y^2 - y\right)^2 = \left(2\sqrt{x^2 + y^2}\right)^2 = 4x^2 + 4y^2;$$
$$x^4 + 2x^2 y^2 + y^4 - 2x^2 y - 2y^3 + y^2 = 4x^2 + 4y^2;$$
$$x^4 + 2x^2 y^2 + y^4 - 2x^2 y - 2y^3 - 4x^2 - 3y^2 = 0.$$

22. $y = x - 2$; $r\left(\cos \theta - \sin \theta\right) = 2$, $-5\pi/4 < \theta < \pi/4$

28. $(x - 5)^2 + (y + 2)^2 = 25$; $r^2 - 10r \cos \theta + 4r \sin \theta + 4 = 0$

34. Matches 10.2.22

40. Multiply each side by r, then complete the square to obtain the equation

$$(x - 1)^2 + (y - 1)^2 = 2.$$

The graph is a circle, center at $(1, 1)$, radius $\sqrt{2}$. It has no symmetries of the sort mentioned; its graph is shown on the right.

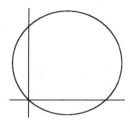

46. Given:
$$r^2 = 4 \cos 2\theta.$$

Note that there is no graph when $\cos 2\theta < 0$; that is, for $\pi/4 < \theta < 3\pi/4$ and for $5\pi/4 < \theta < 7\pi/4$. The graph is symmetric about both coordinate axes and about the origin, and is shown on the right.

52. Given:
$$r^2 = 4 \sin \theta.$$

The graph is symmetric about the x-axis, the y-axis, and the origin. It is shown at the right.

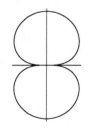

55. The graph of $r = \sin \theta$ is shown as a dashed line in the figure to the right, while the other graph— that of the equation $r = \cos 2\theta$—is shown as a solid line. The Answer section of the text gives the polar coordinates of the four points of intersection.

58. The graph of $r^2 = 4\sin\theta$ is shown as a dashed line in the figure to the right, while the graph of $r^2 = 4\cos\theta$ is a solid line. The five intersection points are the origin and the four points for which $r = 8^{1/4}$ and θ is an odd multiple of $\pi/4$.

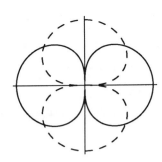

Section 10.3

1. $A = 2\displaystyle\int_0^{\pi/2} \tfrac{1}{2}\left(4\cos^2\theta\right)\, d\theta.$

4. $A = \displaystyle\int_0^{2\pi} \tfrac{1}{2}(4)\left(1 - \sin\theta\right)^2\, d\theta = 2\int_0^{2\pi} \left(1 - 2\sin\theta + \tfrac{1}{2}\left(1 - \cos 2\theta\right)\right)\, d\theta$

$$= \left[2\theta + 4\cos\theta + \theta - \tfrac{1}{2}\sin 2\theta\right]_0^{2\pi} = 6\pi.$$

7. $A = \tfrac{1}{2}\displaystyle\int_0^{\pi} 16\cos^2\theta\, d\theta = 4\int_0^{\pi}\left(1 + \cos 2\theta\right)\, d\theta = 4\left[\theta + \tfrac{1}{2}\sin 2\theta\right]_0^{\pi} = 4\pi.$

10. $A = \tfrac{1}{2}\displaystyle\int_0^{2\pi}\left(10 + 6\sin\theta + 6\cos\theta + 2\sin\theta\,\cos\theta\right)\, d\theta = \tfrac{1}{2}\left[10\theta - 6\cos\theta + 6\sin\theta + \sin^2\theta\right]_0^{2\pi} = 10\pi.$

13. The graph of the given curve is shown on the right. The area within each of the eight loops is

$$A = 2\int_0^{\pi/8} \tfrac{1}{2}\left(4\cos^2 4\theta\right)\, d\theta.$$

16. The graph of $r^2 = 4\cos 2\theta$ is shown on the right. The area within each of the loops is

$$A = \tfrac{1}{2}\int_{-\pi/4}^{\pi/4} 4\cos 2\theta\, d\theta$$

$$= \left[\sin 2\theta\right]_{-\pi/4}^{\pi/4} = 2.$$

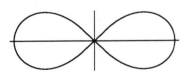

19. The two curves are shown on the right. The region in question has area

$$A = \tfrac{1}{2}\int_{\pi/6}^{5\pi/6}\left(4\sin^2\theta - 1\right)\, d\theta.$$

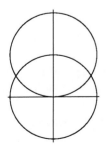

22. We are to find the area outside the circle $r = 2$ and within the curve $r = 2 + \cos\theta$; these curves are shown on the right. The curves meet where $\theta = \pi/2$ and where $\theta = -\pi/2$, so the area in question is

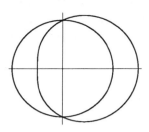

$$A = \tfrac{1}{2} \int_{-\pi/2}^{\pi/2} \left((2 + \cos\theta)^2 - 2^2 \right) d\theta$$

$$= \int_0^{\pi/2} \left(4\cos\theta + \tfrac{1}{2}(1 + \cos 2\theta) \right) d\theta$$

$$= \left[4\sin\theta + \tfrac{1}{2}\theta + \tfrac{1}{4}\sin 2\theta \right]_0^{\pi/2} = \tfrac{1}{4}(\pi + 16).$$

25. The curves, shown on the right, intersect where $\theta = \pi/8$. By symmetry, the area between them is

$$A = 4 \int_0^{\pi/8} \tfrac{1}{2}\sin 2\theta \, d\theta.$$

28. The two curves meet at the origin, at the point with Cartesian coordinates $(-2, 0)$, and at the two points where $\cos\theta = 3 - 2\sqrt{2}$. Let the least positive value of θ that satisfies the latter equation be denoted by ω. Then the area within the figure-8 curve but outside the cardioid is given by

$$A = \int_0^\omega \left(4\cos\theta - (1 - \cos\theta)^2 \right) d\theta = \int_0^\omega \left(6\cos\theta - 1 - \tfrac{1}{2}(1 + \cos 2\theta) \right) d\theta$$

$$= \left[6\sin\theta - \tfrac{3}{2}\theta - \tfrac{1}{4}\sin 2\theta \right]_0^\omega$$

$$= 12\sqrt{3\sqrt{2} - 4} - \tfrac{3}{2}\omega - \left(3 - 2\sqrt{2}\right)\sqrt{3\sqrt{2} - 4} \approx 3.7289587.$$

31. The Cartesian equation of the curve is $\left(x - \tfrac{1}{2}\right)^2 + \left(y - \tfrac{1}{2}\right)^2 = \left(\tfrac{1}{2}\sqrt{2}\right)^2$. So the graph is a circle of radius $\tfrac{1}{2}\sqrt{2}$.

34. One way to solve this problem is to let one circle have polar equation $r = 2a\cos\theta$ and the other have polar equation $r = a$. The first circle then has center at $x = a$ on the x-axis and passes through the origin; the second has center at the origin and passes through the center of the first. They meet where $\theta = \pi/4$ and where $\theta = -\pi/4$, and the area within either that is outside the other is

$$A_1 = \int_0^{\pi/4} \left(4a^2\cos^2\theta - a^2 \right) d\theta = a^2 \int_0^{\pi/4} \left(2(1 + \cos 2\theta) - 1 \right) d\theta = a^2 \left[\theta + \sin 2\theta \right]_0^{\pi/4} = \tfrac{1}{4}a^2(\pi + 4).$$

Let A denote the area within both circles. Each circle has area πa^2, so

$$2\pi a^2 = (A_1 + A) + (A_1 + A). \quad \text{Therefore}$$
$$A = \pi a^2 - A_1 = \tfrac{1}{4}a^2(3\pi - 4).$$

Section 10.4

1. The value of p is 3, so the parabola has equation $y^2 = 12x$.

4. The value of p is 2, so this parabola has equation $(y + 1)^2 = -8(x + 1)$. Its vertex is at the point $(-1, -1)$ and it opens to the left.

7. Here, we have $p = \frac{3}{2}$, so this parabola has equation $x^2 = -6\left(y + \frac{3}{2}\right)$.

10. Here we have $p = 1$, so this parabola has equation

$$(y - 1)^2 = 4(x + 3)$$

and its graph is shown at the right.

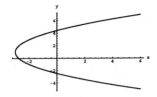

16. The given equation may be written in the form

$$y^2 + 6y + 9 = 2x - 6;$$
$$(y + 3)^2 = (4)(\tfrac{1}{2})(x - 3).$$

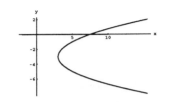

So $p = \frac{1}{2}$; vertex $(3, -3)$; directrix $x = \frac{5}{2}$; focus at $(\frac{7}{2}, -3)$; axis the horizontal line $y = -3$. The graph is on the right.

19. Note that this is a maximum-minimum problem, and recall that a distance is minimized when its square is minimized.

22. Set up coordinates so that the parabola has vertex $V(-p, 0)$. Then the equation of the comet's orbit is

$$y^2 = 4p(x + p).$$

The line $y = x$ meets the orbit of the comet at the point (a, b) (say), which is $100\sqrt{2}$ million miles from the origin (which is also where both the sun and the focus of the parabola are located). Therefore

$$a^2 = 4p(a + p) \quad \text{and} \quad \sqrt{a^2 + a^2} = \left(100\sqrt{2}\right)(10^6) = 10^8\sqrt{2}.$$

It follows that $a = 10^8$. Next, $a^2 = 4p(a + p)$. We apply the quadratic formula to find that $p = \frac{1}{2}\left(\sqrt{2} - 1\right)(10^8)$. The vertex is at distance p from the focus; therefore, by the result of Problem 19, the closest approach is approximately 20.71 million miles.

28. (a) $\alpha = \pi/6$: Range: $\dfrac{2500}{9.8} \cdot \dfrac{\sqrt{3}}{2} \approx 220.9$ meters

Time aloft: $T = \dfrac{2v_0 \sin \alpha}{3} = \dfrac{100(0.5)}{9.8} \approx 5.102$ seconds.

(b) $\alpha = \pi/3$: Range: $\dfrac{2500}{9.8} \cdot \dfrac{\sqrt{3}}{2} \approx 220.9$ meters

Time aloft: $T = \dfrac{100}{9.8} \cdot \dfrac{\sqrt{3}}{2} \approx 8.837$ seconds.

Section 10.5

1. The location of the vertices makes it clear that the center of the ellipse is at $(0, 0)$. Therefore its equation may be written in the standard form

$$\left(\frac{x}{4}\right)^2 + \left(\frac{y}{5}\right)^2 = 1.$$

4. We immediately have $a = 6$ and $b = 4$, so the equation is

$$\left(\frac{x}{4}\right)^2 + \left(\frac{y}{6}\right)^2 = 1.$$

7. First, $a = 10$ and $e = \frac{1}{2}$. So $c = ae = 5$ and thus $b^2 = 100 - 25 = 75$. So the equation is

$$\frac{x^2}{100} + \frac{y^2}{75} = 1.$$

10. First, $c = 4$ and $9 = c/e^2$; therefore $e = \frac{2}{3}$. So $a = c/e = 6$; $b^2 = a^2 - c^2 = 20$. Equation:

$$\frac{x^2}{20} + \frac{y^2}{36} = 1.$$

13. "Move" the center first to the origin to obtain the equation $\dfrac{x^2}{25} + \dfrac{y^2}{16} = 1$. Then apply the translation principle to obtain the answer.

16. In standard form, the equation of this ellipse is

$$\frac{x^2}{4} + \frac{y^2}{16} = 1.$$

So the major axis is vertical, $b = 2$, $a = 4$, the minor axis is of length 4, and the major axis is of length 8. The center is at the origin. Foci: $(0, -2\sqrt{3})$ and $(0, 2\sqrt{3})$. Its graph is shown at the right.

19. The equation can be put in the standard form $\left(\dfrac{x}{2}\right)^2 + \left(\dfrac{y-4}{3}\right)^2 = 1$. The rest is routine.

22. In the usual notation, we have $2a = 0.467 + 0.307 = 0.774$. So $a = 0.387$, $e = 0.206$. Therefore $c = ae = 0.079722$, and

$$b = \sqrt{a^2 - c^2} \approx 0.378699621;$$

we'll use $b = 0.3787$. Therefore the ellipse has major axis 0.774, minor axis 0.7574; in terms of percentages, a is about 2.2% greater than b. Is this a nearly circular orbit? Decide for yourself: Compare the circle (on the left) below, with diameter 0.7657, with the ellipse (on the right) below with the shape of the orbit of the planet.

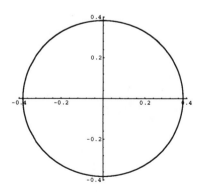
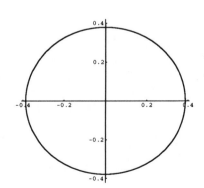

28. Begin with the equation

$$\sqrt{(x-3)^2 + (y+3)^2} + \sqrt{(x+3)^2 + (y-3)^2} = 10.$$

This leads to the equation $16x^2 + 18xy + 16y^2 = 175$.

1. It follows that $c = 4$, $a = 1$, and $b^2 = 15$. So the standard equation is

$$x^2 - \frac{1}{15}y^2 = 1.$$

4. First, $a = 3$, and $b/a = 3/4 = b/3$, so $b = 9/4$. So the equation of the hyperbola is

$$\tfrac{1}{9}x^2 - \tfrac{16}{81}y^2 = 1.$$

7. Let us interchange x and y in the data given in the problem, then restore their meanings with a second interchange at the last step. That is, we assume the foci to be at $(-6, 0)$ and $(6, 0)$, and that the eccentricity is $e = 2$. It follows that $c = 6$, and thus that $a = c/e = 3$. Thus $b^2 = c^2 - a^2 = 27$, and this hyperbola would have equation $x^2/9 - y^2/27 = 1$. Now we interchange x and y to obtain the correct answer.

10. We have $c = 9$; also, $4 = c/e^2$, so $e = 3/2$. Next, $a = c/e = 6$, so $b^2 = c^2 - a^2 = 45$. Hence the equation is

$$\frac{y^2}{36} - \frac{x^2}{45} = 1.$$

13. We "move" the data so that the center is at $(0, 0)$: Replace x by $x - 1$ and y by $y + 2$. The translated hyperbola has vertices at $(0, 3)$ and $(0, -3)$ and asymptotes $3x = 2y$ and $3x = -2y$. Now let's interchange x and y. One asymptote then has equation $y = 2x/3$, so $b/a = 2/3$. But $a = 3$, so $b = 2$. The equation of this hyperbola is $x^2/9 - y^2/4 = 1$. But we must interchange x and y again and then replace y by $y + 2$ and x by $x - 1$; this gives the answer.

16. Write $(x + 2)^2 - 2y^2 = 4$, thus $\dfrac{(x + 2)^2}{4} - \dfrac{y^2}{2} = 1$. So $a = 2$, $b = \sqrt{2}$, and $c^2 = a^2 + b^2 = 6$. Therefore the center is at the point $(-2, 0)$, the foci are at $(-2 + \sqrt{6}, 0)$ and $(-2 - \sqrt{6}, 0)$, and the asymptotes have the equations

$$y = \left| \tfrac{1}{2}\sqrt{2}\,(x + 2) \right|.$$

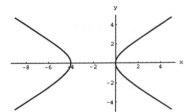

The graph is shown at the right.

22. First show that the equation of the tangent at $P_0(x_0, y_0)$ is

$$y - y_0 = \left(\frac{b^2 x_0}{a^2 y_0} \right)(x - x_0).$$

If $y_0 = 0$, use dx/dy rather than dy/dx to get the desired result.

25. Note that $a^2 = \frac{9}{2} + \frac{9}{2}$, so that $a = 3$; therefore $2a = 6$. Then

$$\sqrt{(x - 5)^2 + (y - 5)^2} + 6 = \sqrt{(x + 5)^2 + (y + 5)^2}.$$

Now apply either the technique or the formula of the solution of Problem 24.

Section 10.7

Note: The following BASIC program was used to verify the solutions of most of the problems of this section. It follows exactly the discussion of rotation of axes given in the text.

```
10  DEFDBL A - Z

15      REM  all variables are double precision.
```

```
20   Z=1.0D-10
25       REM   Criterion for being near zero.
30   P=3.14159265358979324
35       REM   Value for pi.
40   PRINT  "  Enter coefficients A, B, C, D, E, and F"
50   PRINT  "       for the equation"
60   PRINT  "  A*X*X + B*X*Y + C*Y*Y + D*X + E*Y + F = 0."
70   PRINT
80   INPUT  "  A, B, C, D, E, F:  "; A, B, C, D, E, F
90   IF A <> C GOTO 140
95       REM   Go to 140 if the rotation angle is not 45°.
100  L = P/4.0D0
105      REM   This is the case in which the angle is 45°.
110  M =0.5D0*1.414213562D0
115      REM   Half the square root of 2.
120  N = M
130  GOTO 220
135      REM   Computations will be done directly at 220.
140  L = 0.5D0*ATN(B/(A — C))
145        Rotation angle in general case.
150  IF L >=0.0D0 GOTO 170
155      REM   Jump to 170 if the angle is nonnegative.
160  L = L + 0.5D0*P
165      REM   Correction of negative angle.
170  O = ABS(A — C)
180  Y=O/SQR(O*O + B*B)
190  Y=Y*SGN((A — C)/B
195      REM   Note assumption:   B is nonzero.
200  M = SQR(0.5D0*(1.0D0 — Y))
210  N = SQR(0.5D0*(1.0D0 + Y))
215      REM   Equations from the text appear next.
220  G = A*N*N + B*N*M + C*M*M
230  H = B*(N*N — M*M) + 2.0D0*(C — A)*N*M
240  I = A*M*M — B*M*N + C*N*N
250  J = D*N + E*M
260  K = E*N — D*M
270  Q = B*B — 4.0D0*A*C
275      REM   Note the discriminant in line 270.
280  IF Q<Z GOTO 310
285      REM   Otherwise Q is surely positive.
```

```
290  PRINT  "        Hyperbola:"
300  GOTO 360
305      REM  Line 360 is where we print the parameters.
310  If Q > -Z GOTO 340
315      REM Go to line 340 because Q is probably nonnegative.
320  PRINT  "        Ellipse:"
325      REM  Because Q is definitely negative.
330  GOTO 360
335      REM To print parameters.
340  PRINT  "        Parabola"
350  PRINT  "                ... probably.  Discriminant:    "; Q
355      REM  Because Q is zero or nearly zero.
360  PRINT  " Rotation angle:   Alpha = "; L
370  R = 180.0D0*L/P
375      REM  Conversion to degrees.
380  PRINT  " In degrees, that's "; R
390  PRINT  "      Its sine is "; M
395      REM To help identify the rotation angle.
400  PRINT  "      Its cosine is "; N
410  S = M/N
420  PRINT  "      Its tangent is "; S
430  PRINT  " In the appropriately rotated uv-coordinate system,"
440  PRINT  "    the equation of the conic takes the form"
450  PRINT
460  PRINT  "    "; G;  "*U*U + ";  H;  "*U*V"
470  PRINT  "        + ";  I;  "*V*V + ";  J;  "*U"
480  PRINT  "        + ";  K;  "*V + ";  F;  "    =   0."
490  PRINT
500  END
```

---

4. $(y - 2)^2 = 4(x + 3)$.   Parabola: $(y')^2 = 4x'$; origin $(2, -3)$.

10. $10(x')^2 + 5(y')^2 = 40$.   Ellipse; $\theta = \arctan\left(\frac{1}{2}\right) \approx 26.57°$.

16. $25(x')^2 = 0$.   The line $x' = 0$; $\theta = \arctan\left(\frac{4}{3}\right) \approx 53.13°$.

22. $24(x' - 2)^2 + 50(y' - 1)^2 = 25$. Ellipse; $\theta = \arctan\left(\frac{4}{3}\right) \approx 53.13°$.

31. We require α so that $A' = 0$. We have $A = 1$ and $C = -1$, and the coefficients B, D, E, and F are fixed but arbitrary.

For $A' = 0$: $\cos^2 \alpha + B \cos \alpha \sin \alpha - \sin^2 \alpha = 0$;
For $C' = 0$: $\sin^2 \alpha - B \sin \alpha \cos \alpha - \cos^2 \alpha = 0$.

These equations imply that $\cos 2\alpha + \frac{1}{2} B \sin 2\alpha = 0$, and it follows that $\cot 2\alpha = -\dfrac{B}{2}$.

Now the cotangent function takes on all real values on the open interval $(0, \pi/2)$, so there is always a unique first-quadrant angle α such that $\cot 2\alpha = -B/2$. The transformed equation has the form

$$B'x'y' + D'x' + E'y' + F' = 0,$$

and this equation may be written in the form

$$y' = -\frac{D'x' + F'}{B'x' + E'}.$$

As $x' \to +\infty$, $y' \to -D/B'$. If you verify that $B' = -\sqrt{B^2 + 4} < 0$, you may conclude that (except in degenerate cases) the graph is a hyperbola with an asymptote at the angle α with the x-axis.

34. Given: $x^2 + 14xy + 49y^2 = 100$. In the appropriate rotated uv-coordinate system, this equation becomes

$$50u^2 = 100; \quad \text{that is,} \quad u = \sqrt{2} \text{ or } u = -\sqrt{2}.$$

So its graph is a pair of parallel lines—a degenerate parabola. The angle of rotation of the graph is $\tan^{-1}(7) \approx 81°52'12''$. The closest approach to the origin occurs when $|u| = \sqrt{2}$ and $v = 0$, in which case the distance to the origin is $\sqrt{2}$. The two points of this degenerate parabola closest to the origin have xy-coordinates $\left(\frac{1}{5}, \frac{7}{5}\right)$ and $\left(-\frac{1}{5}, -\frac{7}{5}\right)$.

Chapter 10 Miscellaneous

4. Given: $y^2 = 4(x + y)$. Thus

$$y^2 - 4y + 4 = 4x + 4 = 4(x + 1);$$

it follows that

$$(y - 2)^2 = 4(x + 1).$$

So the graph is a parabola; it has vertex at $(-1, 2)$, directrix $x = -2$, axis $y = 2$, focus at $(0, 2)$, and it opens to the right. Its graph appears at the right.

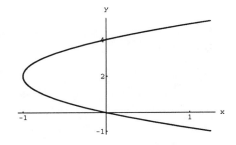

10. The given equation can be written in the form

$$\frac{(y - 1)^2}{4} - \frac{(x + 1)^2}{9} = 1.$$

So the graph is a hyperbola with center $(-1, 1)$. Because $c = \sqrt{13}$, the foci are at the points

$$(-1, 1 + \sqrt{13}) \quad \text{and} \quad (-1, 1 - \sqrt{13}).$$

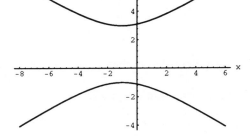

The vertices are at $(-1, 3)$ and $(-1, -1)$. The transverse axis is vertical, of length 4, and the conjugate axis is horizontal, of length 6. The eccentricity is $\frac{1}{2}\sqrt{13}$, the directrices are

$$y = 1 + \tfrac{4}{13}\sqrt{13} \text{ and } y = 1 - \tfrac{4}{13}\sqrt{13},$$

and the asymptotes have the equations $3y = 2x + 5$ and $3y = -2x + 1$. The graph is shown above.

16. In a *uv*-coordinate system rotated 45° from the *xy*-system, the given equation takes the form

$$v^2 - \tfrac{1}{2}u^2 = 1.$$

This is the equation of a hyperbola with center $(0,0)$, vertices $(0,1)$ and $(0,-1)$; the foci are at $(0,\sqrt{3})$ and $(0,-\sqrt{3})$. The transverse axis is parallel to the *v*-axis and has length 2, the conjugate axis has length $2\sqrt{2}$. The eccentricity is $e = \sqrt{3}$, the directrices satisfy

$$|v| = \tfrac{1}{3}\sqrt{3},$$

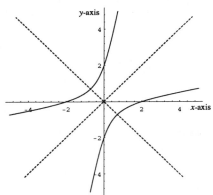

and the asymptotes satisfy $|v| = \tfrac{1}{2}|u|\sqrt{2}$. The graph is shown above.

22. Multiply each side by r to obtain $y^2 = x$. This is the equation of a parabola with axis the *x*-axis, opening to the right, with vertex $(0,0)$, directrix $x = -\tfrac{1}{4}$, and focus $(\tfrac{1}{4},0)$.

28. Given: $r = \dfrac{1}{1 + \cos\theta}$. In Cartesian coordinates we obtain $y^2 = -2x + 1$, so it is a parabola with focus at $(0,0)$, directrix $x = 1$, and the vertex is at $(0,\tfrac{1}{2})$. It opens to the left and its axis is the *x*-axis.

31. The region whose area is sought is shown at the right.

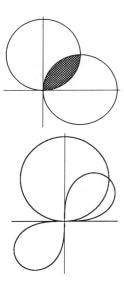

34. The two regions are shown at the right. The curves cross at the points where $\sin\theta = 0$ and where $\cos\theta = \sin\theta$. We obtain the two solutions $r = 0$ and $r = \sqrt{2}$, $\theta = \pi/4$. The area of the small region is

$$A_S = \tfrac{1}{2}\int_0^{\pi/4} \left((2\sin 2\theta) - (2\sin\theta)^2\right)\, d\theta,$$

which turns out to be $\tfrac{1}{4}(4 - \pi) \approx 0.214602$. The area of the large region is

$$A_L = \int_0^{\pi/2} \sin 2\theta\, d\theta = 1.$$

Therefore the total area outside the circle but within the lemniscate is $\tfrac{1}{4}(8 - \pi) \approx 1.214602$.

37. The circle and the cardioid are shown at the right. They intersect only at the pole.

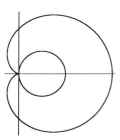

40. From the figure at the right, we see that matters would be greatly simplified if we were to rotate 45° counterclockwise to obtain the situation shown in the following figure. Then the parabola in the second figure has equation

$$r = \frac{2^{3/2}}{1 - \cos\theta}.$$

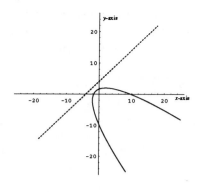

Therefore the parabola in the first figure has equation

$$r = \frac{2^{3/2}}{1 - \cos(\theta + (\pi/4))}.$$

After a considerable amount of algebra we find that we can write the Cartesian equation of the first parabola in the form

$$x^2 + 2xy + y^2 - 8x + 8y - 16 = 0.$$

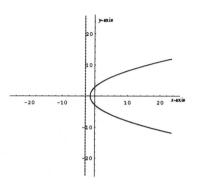

46. We use the figure below. Note that we introduce a uv-coordinate system; in the rest of this discussion, all coordinates will be uv-coordinates.

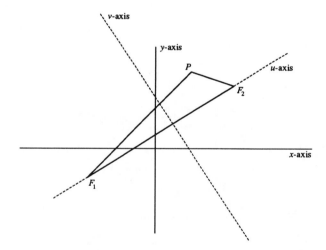

Choose the new axes so that $F_1 = F_1(c, 0)$ and $F_2 = F_2(-c, 0)$, $c > 0$. Suppose that $P = P(u, v)$. Then

$$|PF_2| = 2a + |PF_1|,$$

and therefore

$$\sqrt{(u + c)^2 + v^2} = 2a + \sqrt{(u - c)^2 + v^2}.$$

Consequently

$$(u + c)^2 + v^2 = 4a^2 + 4a\sqrt{(u - c)^2 + v^2} + (u - c)^2 + v^2.$$

Successive simplifications produce

$$4uc - 4a^2 = 4a\sqrt{(u - c)^2 + v^2};$$

$$uc - a^2 = a\sqrt{(u - c)^2 + v^2};$$

$$u^2c^2 - 2a^2uc + a^4 = a^2u^2 - 2a^2uc + a^2c^2 + a^2v^2;$$

$$u^2c^2 - a^2u^2 - a^2v^2 = a^2c^2 - a^4;$$

$$u^2(c^2 - a^2) - a^2v^2 = a^2(c^2 - a^2).$$

Now $|F_1 F_2| > 2a$, so $c > a$. Thus $c^2 - a^2 = b^2$ for some $b > 0$. Hence

$$b^2u^2 - a^2v^2 = a^2b^2; \quad \text{that is,}$$

$$\frac{u^2}{a^2} - \frac{v^2}{b^2} = 1.$$

Therefore the locus of $P(u, v)$ is a hyperbola with vertices $(a, 0)$ and $(-a, 0)$ and foci $(c, 0)$ and $(-c, 0)$ (because $c^2 = a^2 + b^2$), and therefore the hyperbola has foci F_1 and F_2. Finally, if a circle with radius r_2 is centered at F_2 and another with radius r_1 is centered at F_1, with r_2 and r_1 satisfying the equation $r_2 = 2a + r_1$, then the two circles will intersect at a point on the hyperbola. You may thereby construct by straightedge-and-compass methods as many points lying on the hyperbola as you please.

49. The equation may be written in the form

$$(x + 2)^2 - \tfrac{1}{3}y^2 = 1,$$

so the graph is a hyperbola with a vertical directrix. In the notation at the beginning of Section 10.6, we have $a = 1$ and $b = \sqrt{3}$, so $c^2 = a^2 + b^2 = 4$, and therefore $c = 2$. The eccentricity is then $e = c/a = 2$.

52. Here, the loop has polar equation

$$r = \frac{5\cos^2\theta \sin^2\theta}{\cos^5\theta + \sin^5\theta} \quad \text{for } 0 \le \theta \le \pi/2.$$

Therefore the area it bounds is

$$A = 25 \int_0^{\pi/4} \frac{\cos^4\theta \sin^4\theta}{\left(\cos^5\theta + \sin^5\theta\right)^2} \, d\theta.$$

The substitution $u = \tan\theta$ transforms this integral into

$$A = 25 \int_0^1 \frac{u^4}{(1 + u^5)^2} \, du = \tfrac{5}{2}.$$

55. The equation of the conic can be written

$$x^2 + Bxy + Cy^2 + Dx + Ey + F = 0.$$

So $25 + 5D + F = 0$,

$25 - 5D + F = 0$. So $D = 0$ and $F = -25$.

Also $16C + 4E + F = 0$,

$16C - 4E + F = 0$. So $E = 0$ and $F = -16C$.

So $16C = 25$: $C = \frac{25}{16}$. The equation of the conic is thus

$$x^2 + Bxy + \tfrac{25}{16}y^2 - 25 = 0;$$
$$16x^2 + 16By + 25y^2 = 400.$$

The discriminant is $256B^2 - 1600 = 64\left(4B^2 - 25\right)$.

If $B < \frac{5}{2}$, the conic is an ellipse.

If $B = \frac{5}{2}$, the conic has equation $16x^2 + 40y + 25y^2 = 400$;
$$\left(4x + 5y\right)^2 = 400;$$
$$4x + 5y = \pm 20;$$
$$y = -\tfrac{4}{5}x \pm 20:\ \text{two parallel lines.}$$

If $B < \frac{5}{2}$, the conic is a hyperbola.

Chapter 11: Infinite Series

Section 11.2

1. $a_n = \dfrac{2n}{5n-3} = \dfrac{2}{5 - \dfrac{3}{n}} \to \dfrac{2}{5}$ as $n \to +\infty$.

4. $\displaystyle\lim_{n\to\infty} \dfrac{n}{10 + (1/n^2)}$ does not exist.

10. $-1 \le \sin n \le 1$ for all n, so by the squeeze law, $\displaystyle\lim_{n\to\infty} \dfrac{\sin n}{3^n} = 0$.

16. $\tfrac{1}{2}, 2, \tfrac{1}{2}, 2, \ldots$ diverges.

22. Apply l'Hôpital's rule thrice to $\displaystyle\lim_{x\to\infty} \dfrac{x^3}{e^{x/10}}$ to get the limit 0.

28. $\ln a_n = -\dfrac{1}{n} \ln (0.001) = \dfrac{1}{n} \ln (1000) \to 0$ as $n \to \infty$.

31. $\ln a_n = \dfrac{3}{n} \ln \left(\dfrac{2}{n}\right) = \dfrac{3 (\ln 2 - \ln n)}{n} \to 0$.

34. $\displaystyle\lim_{n\to\infty} a_n = \lim_{x\to\infty} \dfrac{(2/3)^x}{1 - x^{1/x}} = \lim_{x\to\infty} \dfrac{x^2 (2/3)^x \ln (2/3)}{((\ln x) - 1) x^{1/x}}$ by l'Hôpital's rule. And

$$\lim_{x\to\infty} x^2 (2/3)^x = \lim_{x\to\infty} \dfrac{x^2}{(3/2)^x}$$
$$= \lim_{x\to\infty} \dfrac{2x}{(3/2)^x \ln (3/2)}$$
$$= \lim_{x\to\infty} \dfrac{2}{(3/2)^x (\ln (3/2))^2} = 0.$$

Thus $\displaystyle\lim_{x\to\infty} \dfrac{x^2 (2/3)^x \ln (2/3)}{((\ln x) - 1) x^{1/x}} = 0$, and therefore $\displaystyle\lim_{n\to\infty} \dfrac{(2/3)^n}{1 - n^{1/n}} = 0$.

40. $\tau = \displaystyle\lim_{n\to\infty} \dfrac{F_{n+1}}{F_n} = \lim_{n\to\infty} \dfrac{F_n + F_{n-1}}{F_n} = \lim_{n\to\infty} \left(1 + \dfrac{F_{n-1}}{F_n}\right) = 1 + \dfrac{1}{\displaystyle\lim_{n\to\infty} \dfrac{F_n}{F_{n-1}}} = 1 + \dfrac{1}{\tau}$.

Therefore $\tau^2 = \tau + 1$.

Section 11.3

1. Geometric, ratio $\tfrac{1}{3}$, first term 1; sum $\dfrac{1}{1 - \tfrac{1}{3}} = \tfrac{3}{2}$.

4. $\displaystyle\lim_{n\to\infty} (0.5)^{1/n} = 1 \ne 0$: This series diverges by the nth term test.

7. Geometric, ratio $\tfrac{1}{3}$.

10. $\displaystyle\lim_{x\to\infty} x^{1/x} = 1 \ne 0$. This series diverges by the nth term test.

13. Geometric, ratio $r = -\dfrac{3}{e}$. The series diverges because $|r| > 1$.

16. $\displaystyle\sum_{n=0}^{\infty} \dfrac{1}{2^n} = 1$. So $\displaystyle\sum_{n=1}^{k} \left(\dfrac{2}{n} - \dfrac{1}{2^n}\right) > \left(\sum_{n=1}^{k} \dfrac{2}{n}\right) - 1$ for all positive integers k.

Because the harmonic series diverges to $+\infty$, so does $\displaystyle\sum_{n=1}^{\infty} \dfrac{2}{n}$, and therefore so does the given series.

19. $\dfrac{\tfrac{1}{5}}{1 - \tfrac{1}{5}} - \dfrac{\tfrac{1}{7}}{1 - \tfrac{1}{7}} = \tfrac{1}{4} - \tfrac{1}{6} = \tfrac{1}{12}$.

22. The series is geometric with ratio $r = \dfrac{\pi}{e}$, and it diverges because $|r| > 1$.

28. Diverges by the nth term test ($2^{1/n} \to 1$ as $n \to +\infty$).

34. $\dfrac{307}{909}$

40. $\dfrac{1}{n^2 - 1} = \dfrac{\frac{1}{2}}{n - 1} + \dfrac{-\frac{1}{2}}{n + 1}$, so

$$
\begin{aligned}
S_k &= \sum_{n=2}^{k} \frac{1}{n^2 - 1} = \frac{1}{2} \sum_{n=2}^{k} \left(\frac{1}{n - 1} - \frac{1}{n + 1} \right) \\
&= \frac{1}{2} \left(1 - \frac{1}{3} + \frac{1}{2} - \frac{1}{4} + \frac{1}{3} - \frac{1}{5} + \frac{1}{4} - \frac{1}{6} + \cdots + \frac{1}{k - 2} - \frac{1}{k} + \frac{1}{k - 1} - \frac{1}{k + 1} \right) \\
&= \tfrac{1}{2} \left(1 + \frac{1}{2} - \frac{1}{k} - \frac{1}{k + 1} \right) \to \frac{3}{4} \quad \text{as} \quad k \to \infty.
\end{aligned}
$$

Therefore $\displaystyle\sum_{n=2}^{\infty} \frac{1}{n^2 - 1} = \frac{3}{4}$.

46. Total distance traveled:

$$
\begin{aligned}
D &= a + 2ra + 2r^2 a + 2r^3 a + 2r^4 a + \ldots \\
&= -a + 2a \left(1 + r + r^2 + r^3 + r^4 + \ldots \right) \\
&= -a + \frac{2a}{1 - r} = \frac{-a + ar + 2a}{1 - r} = a \frac{1 + r}{1 - r}.
\end{aligned}
$$

49. $M_1 = (0.95)\, M_0$, $M_2 = (0.95)\, M_1 = (0.95)^2\, M_0$, and so on. Therefore $M_n = (0.95)^n\, M_0$.

52. Let X denote 1, 2, 3, 4, or 5. Peter (who goes first) has these winning patterns and their probabilities:

6	$\dfrac{1}{6}$
X X X 6	$\dfrac{5^3}{6^4}$
X X X X X X 6	$\dfrac{5^6}{6^7}$
X X X X X X X X X 6	$\dfrac{5^9}{6^{10}}$
$\vdots$	$\vdots$

So his probability of winning the game is the sum of the geometric series with first term $\frac{1}{6}$ and ratio $\frac{125}{216}$, and so his probability of winning the game is $\frac{36}{91} \approx 0.3956$. Paul (who is second) wins with probability $\frac{30}{91} \approx 0.3297$ and Mary wins with probability $\frac{25}{91} \approx 0.2747$.

Section 11.4

4. $\dfrac{1}{1 - x} = 1 + x + x^2 + x^3 + x^4 + \dfrac{x^5}{(1 - z)^6}$ for some z between 0 and x.

10. $f(x) = -7 + 5x - 3x^2 + x^3$.

16. $\tan x = 1 + \dfrac{2}{1!} \left(x - \dfrac{\pi}{4} \right) + \dfrac{4}{2!} \left(x - \dfrac{\pi}{4} \right)^2 + \dfrac{16}{3!} \left(x - \dfrac{\pi}{4} \right)^3 + \dfrac{80}{4!} \left(x - \dfrac{\pi}{4} \right)^4 + \dfrac{Q(z)}{5!} \left(x - \dfrac{\pi}{4} \right)^5$ for some z between $\pi/4$ and x, where $Q(z) = 16 \sec^6 z + 88 \sec^4 z \tan^2 z + 16 \sec^2 z \tan^4 z$.

22. $1 + 2x + \dfrac{1}{2!} 4x^2 + \dfrac{1}{3!} 8x^3 + \dfrac{1}{3!} 16x^4 + \ldots$

25. $1 - \frac{1}{2!} 4x^2 + \frac{1}{4!} 16x^4 - \frac{1}{6!} 64x^6 + \frac{1}{8!} 256x^8 - \ldots$

28. $1 - \frac{1}{2!} x + \frac{1}{4!} x^2 - \frac{1}{6!} x^3 + \frac{1}{8!} x^4 - \ldots$

34. $1 - \frac{1}{2!} \left(x - \frac{\pi}{2}\right)^2 + \frac{1}{4!} \left(x - \frac{\pi}{2}\right)^4 - \frac{1}{6!} \left(x - \frac{\pi}{2}\right)^6 + \ldots$

40. $-\left(x - \frac{\pi}{2}\right) + \frac{1}{3!} \left(x - \frac{\pi}{2}\right)^3 - \frac{1}{5!} \left(x - \frac{\pi}{2}\right)^5 + \ldots$

46. Given: $\alpha = \tan^{-1}(1/5)$.

(a): $\tan 2\alpha = \dfrac{\frac{1}{5} + \frac{1}{5}}{1 - \frac{1}{25}} = \frac{5}{12}$.

(b): $\tan 4\alpha = \dfrac{\frac{5}{12} + \frac{5}{12}}{1 - \frac{25}{144}} = \frac{120}{119}$.

(c): $\tan\left(\dfrac{\pi}{4} - 4\alpha\right) = \dfrac{1 - \frac{120}{119}}{1 + \frac{120}{119}} = -\frac{1}{239}$.

(d): $\tan\left(\dfrac{\pi}{4} - 4\alpha\right) = -\frac{1}{239}$:

$\dfrac{\pi}{4} - 4\alpha = -\arctan \frac{1}{239}$;

$4\arctan \frac{1}{5} - \arctan \frac{1}{239} = \dfrac{\pi}{4}$.

Section 11.5

1. $\displaystyle\int_0^\infty \frac{x}{x^2 + 1}\, dx = \left[\frac{1}{2} \ln(x^2 + 1)\right]_0^\infty = +\infty$. Therefore the given series diverges.

4. $\displaystyle\int_0^\infty (x + 1)^{-4/3}\, dx = \left[-3(x + 1)^{-1/3}\right]_0^\infty = 3 < \infty$. Therefore the given series converges.

7. $\displaystyle\int_2^\infty \frac{1}{x \ln x}\, dx = \left[\ln(\ln x)\right]_2^\infty = +\infty$.

10. $\displaystyle\int_0^\infty x e^{-x}\, dx = 1 < \infty$, so the given series converges.

13. $\displaystyle\int_1^\infty \frac{\ln x}{x^2}\, dx = \left[-\frac{1}{x}(1 + \ln x)\right]_1^\infty = +\infty$. (Integrate by parts; use l'Hôpital's rule to get the limit.)

16. $\displaystyle\int_1^\infty \frac{dx}{x^3 + x} = \int_1^\infty \left(\frac{1}{x} - \frac{x}{x^2 + 1}\right) dx = \left[(\ln x) - \frac{1}{2}\ln(x^2 + 1)\right]_1^\infty = \frac{1}{2}\left[\ln \frac{x^2}{x^2 + 1}\right]_1^\infty = \frac{1}{2}\ln 2 < \infty$,

and therefore the given series converges.

19. Let $J = \displaystyle\int_1^\infty \ln\left(1 + \frac{1}{x^2}\right) dx$. Integrate by parts: With $u = \ln(1 + x^{-2})$ and $dv = dx$, we find that

$$J = \left[x \ln(1 + x^{-2})\right]_1^\infty + 2\int_1^\infty \frac{dx}{1 + x^2}$$

$$= \lim_{x \to \infty} \frac{\ln(1 + x^{-2})}{1/x} - \ln 2 + \left[2 \tan^{-1} x\right]_1^\infty$$

$$= \left(\lim_{u \to 0^+} \frac{\ln(1 + u^2)}{u}\right) + \pi - \frac{\pi}{2} - \ln 2 = \frac{\pi}{2} - \ln 2 < \infty,$$

so the series converges.

22. $\displaystyle\int_1^\infty \frac{x}{(4x^2 + 5)^{3/2}}\, dx = \left[-\frac{1}{4\sqrt{4x^2 + 5}}\right]_1^\infty < \infty$. Therefore the given series converges.

28. $\displaystyle\int_1^\infty (x + 1)^{-3}\, dx = \left[-\frac{1}{2}(x + 1)^{-2}\right]_1^\infty = \frac{1}{8} < \infty$. Therefore the given series converges.

34. We require $R_n < (2)\left(10^{-11}\right)$. This will hold provided that

$$\int_n^\infty \frac{1}{x^6}\,dx < (2)\left(10^{-11}\right)$$

because R_n cannot exceed the integral by Theorem 2. So we require

$$\left[-\frac{1}{5x^5}\right]_n^\infty < (2)\left(10^{-11}\right),$$

and it follows that $n^5 > 10^{10}$, and thus that $n > 100$. We therefore choose $n = 101$. The exact value of the sum is $\dfrac{\pi^6}{945} \approx 1.017343061984$.

37. Let $S = \displaystyle\sum_{n=1}^\infty \frac{1}{n^5}$. Then $S - S_n = a_{n+1} + a_{n+2} + \ \dots\ = R_n = \displaystyle\sum_{k=n+1}^\infty \frac{1}{k^5}$.

$$\int_{n+1}^\infty x^{-5}\,dx \le R_n \le \int_n^\infty x^{-5}\,dx :$$
$$\left[\frac{1}{4n^4}\right]_\infty^{n+1} \le R_n \le \left[\frac{1}{4x^4}\right]_\infty^{n} ;$$
$$\frac{1}{4\left(n+1\right)^4} \le R_n \le \frac{1}{4n^4}.$$

For the accuracy required, we need to have $\dfrac{1}{4n^4} - \dfrac{1}{4\left(n+1\right)^4} < 0.0005$, and to achieve these ends we choose $n = 6$. Then

$$S_6 + \frac{1}{(4)\left(7^4\right)} \le S \le S_6 + \frac{1}{(4)\left(6^4\right)};$$
$$1.036790390 + 0.000104123 \le S \le 1.036790390 + 0.000192901, \quad \text{so that}$$
$$1.036894513 \le S \le 1.036983291.$$

So we can definitely say that $1.03689 \le S \le 1.03699$. The sum is approximately 1.03692775514337.

40. Show that $c_n - c_{n+1} = \displaystyle\int_n^{n+1} \frac{1}{x}\,dx - \frac{1}{n+1} > 0$. A better approximation is $\gamma \approx 0.577215664902$.

Section 11.6

1. Because $n^2 + n + 1 > n^2$ if $n > 0$, this series converges by comparison with the p-series for which $p = 2$.

4. Converges by limit-comparison with the p-series for which $p = \frac{3}{2}$.

7. Diverges by limit-comparison with the harmonic series.

10. Diverges by limit-comparison with the harmonic series.

13. Diverges by *careful* limit-comparison with the harmonic series.

16. Converges by comparison with the geometric series with ratio $\frac{1}{3}$.

19. Because $n^2 \ln n > n^2$ if $n \ge 3$, this series converges by comparison with the p-series for which $p = 2$.

22. Diverges by limit-comparison with the harmonic series.

25. Converges by comparison with the p-series for which $p = 2$. To show this, show that $\ln n < n$ if $n > 2$ and that $e^n > n^3$ if $n > 4$. To do the latter, sketch the graph of $f(x) = e^x - x^3$. Obtain an estimate of its largest x-intercept (it's approximately 4.5364) and show that the graph of $f(x)$ is increasing for all larger values of x.

28. Converges by comparison with the geometric series with ratio $\frac{1}{2}$.

34. Show that $\ln n < n$ if $n \geq 1$. Then the given series converges by comparison with the p-series for which $p = 2$.

40. First, $1 + 2 + 3 + 4 + \ \ldots \ + n = \dfrac{n(n+1)}{2}$, and so $\dfrac{1}{1 + 2 + 3 + \ \ldots \ + n} = \dfrac{2}{n(n+1)} = \dfrac{2}{n} - \dfrac{2}{n+1}$. So the sum of the first k terms of the given series is

$$S_k = 2\left(1 - \frac{1}{k+1}\right).$$

Therefore the sum of the series is 2.

Section 11.7

1. Converges by Theorem 1. Note: The sum of this series is $\dfrac{\pi^2}{12}$.

4. Diverges by the nth term test.

7. Converges by Theorem 1. Note: The sum of this series is $\dfrac{1}{e}$.

10. Converges by the ratio test.

13. Converges conditionally—Theorem 1 for convergence, a careful limit-comparison with the harmonic series to show that the given series is not absolutely convergent.

16. Converges absolutely by the ratio test.

19. Diverges by the nth term test.

22. Diverges by the nth term test.

25. Diverges by the nth term test: $\ln(1/n) = -\ln n$.

28. Claim: $f(x) = \dfrac{1}{x}\arctan x$ is decreasing for $x \geq 2$. The reason is that

$$f'(x) = \frac{x - (1 + x^2)\arctan x}{x^2(1 + x^2)},$$

$$\text{so} \ \ f'(x) < \frac{x - (1 + x^2)}{x^2(1 + x^2)} \ \ \text{if} \ \ x \geq 2;$$

$$\text{thus for} \ \ x \geq 2, \ f'(x) < \frac{2x - (1 + x^2)}{x^2(1 + x^2)} = -\frac{(x - 1)^2}{x^2(x^2 + 1)} < 0.$$

Therefore $f(x) = \dfrac{1}{x}\arctan x$ is decreasing for $x \geq 2$, so the given series converges by Theorem 1. But the convergence is conditional because

$$\sum_{n=1}^{\infty} \frac{1}{n}\arctan n$$

diverges by limit-comparison with the harmonic series.

34. Because $|R_n| < \dfrac{1}{(n+1)^2} < 0.0005$ when $n > 43.72$, we choose $n = 44$.
Numerical notes: $S_{44} \approx 0.8222146356$, $1/(45^2) \approx 0.000493827$, and the sum of the series itself is about 0.822357033. (The exact value of the sum is $\pi^2/12$. You can deduce this from the results of Problem 42 in Section 11.8.)

40. Choose $n \geq 5$; $S_4 \approx 0.971888644$; the sum of the given series is approximately 0.972119771. Answer: 0.972.

43. Let $a_n = b_n = \dfrac{(-1)^{n+1}}{\sqrt{n}}$.

Section 11.8

1. The ratio test yields the limit $|x|$, so the radius of convergence is 1. The series has center zero, converges at $x = -1$ (by the alternating series test), and diverges when $x = 1$ (for then it is the harmonic series). Consequently the interval of convergence is $-1 \leq x < 1$.

4. The interval of convergence is the single point $x = 0$.

10. The ratio test leads to the limit $|x| \left(\displaystyle\lim_{n \to \infty} \left(\dfrac{n}{n+1} \right)^n \right) = \dfrac{|x|}{e}$. So the series converges on the interval $(-e, e)$ and diverges if $|x| > e$. Only if we borrow the result of Miscellaneous Problem 61 of this chapter is it easy to test convergence at the endpoints $-e$ and e: For large values of n,

$$\frac{n!}{n^n} e^n \approx \frac{\sqrt{2\pi n}\,(n/e)^n}{n^n} e^n = \sqrt{2\pi n} \to +\infty,$$

so this series diverges at the end points; the interval of convergence really is $(-e, e)$.

16. $(-\infty, +\infty)$

22. $\dfrac{1}{10} - \dfrac{x}{100} + \dfrac{x^2}{1000} - \dfrac{x^3}{10000} + \dots$; radius 10.

28. $\left(9 + x^3 \right)^{-1/2} = \displaystyle\sum_{n=0}^{\infty} \dfrac{(-1)^n (2n)!\, x^{3n}}{(n!)^2\, 2^{2n} 3^{2n+1}}$; the ratio test gives limit $\frac{1}{9}|x|^3$; the radius of convergence is therefore $9^{1/3}$.

31. Integrate $t^3 - \dfrac{1}{3!} t^9 + \dfrac{1}{5!} t^{15} - \dfrac{1}{7!} t^{21} + \dots$.

34. $f(x) = x - \dfrac{x^3}{3^2} + \dfrac{x^5}{5^2} - \dfrac{x^7}{7^2} + \dfrac{x^9}{9^2} - \dots$; the interval of convergence is $-1 \leq x \leq 1$.

40. $f(x) = \displaystyle\sum_{n=0}^{\infty} a_n x^n$: $f(0) = a_0 = \dfrac{f^{(0)}(0)}{0!}$.

$f'(x) = \displaystyle\sum_{n=1}^{\infty} n a_n x^{n-1}$: $f'(0) = a_1 = \dfrac{f^{(1)}(0)}{1!}$, and so on.

Section 11.9

1. Use the binomial series for $4 (x+1)^{1/3}$; take $x = \frac{1}{64}$.

4. We use the series
$$e^{-x} = 1 - \frac{1}{1!} x + \frac{1}{2!} x^2 - \frac{1}{3!} x^3 + \dots \ .$$

To insure that $|R_n| < 0.0005$, we impose the condition that
$$\frac{(0.2)^{n+1}}{(n+1)!} < \frac{1}{2000},$$

and it suffices to take $n = 3$. We find that
$$e^{-0.2} \approx 1 - \frac{0.2}{1!} + \frac{(0.2)^2}{2!} - \frac{(0.2)^3}{3!} \approx 0.818666667.$$

Answer: 0.819. (The true value is approximately 0.818730753.)

10. First,
$$\cos x = 1 - \frac{1}{2!} x^2 + \frac{1}{4!} x^4 - \dots \ ;$$

the error term in using the Taylor polynomial $P_n(x)$ to approximate $\cos x$ cannot exceed

$$\frac{|x|^{n+1}}{(n+1)!},$$

and here we have $x = \dfrac{\pi}{36} < 0.0873$. For error less than 0.0005, we need $n \geq 2$; we take $n = 2$, and the error will not exceed 0.000111. Result: $\cos(\pi/36) \approx 0.9961922823$. (The true value is approximately 0.9961946981.)

13. Integration produces the series

$$\frac{1}{2} - \frac{1}{3^2 \cdot 2^3} + \frac{1}{5^2 \cdot 2^5} - \frac{1}{7^2 \cdot 2^7} + \cdots .$$

16. For $-1 \leq x \leq 1$, we have

$$\left(1 + x^4\right)^{-1/2} = 1 - \frac{1}{2}x^4 + \frac{(1)(3)}{2! \, 2^2} x^8 - \frac{(1)(3)(5)}{3! \, 2^3} x^{12} + \frac{(1)(3)(5)(7)}{4! \, 2^4} x^{16} - \cdots$$

$$= \sum_{n=0}^{\infty} \frac{(-1)^n (2n)!}{(n!)^2 \, 2^{2n}} x^{4n}.$$

Convergence at the endpoints -1 and 1 follow from an application of Formula 113 of the endpapers of the text and the alternating series test, but all we need here is absolute convergence for $0 \leq x \leq \frac{1}{2}$. Hence, for such x,

$$1 - \frac{1}{2}x^4 + \frac{(1)(3)}{2! \, 2^2} x^8 - \frac{(1)(3)(5)}{3! \, 2^3} x^{12} \leq \frac{1}{\sqrt{1 + x^4}} \leq 1 - \frac{1}{2}x^4 + \frac{(1)(3)}{2! \, 2^2} x^8.$$

When we integrate these terms from $x = 0$ to $x = \frac{1}{2}$, we find that the value of the given integral is trapped between 0.49695344 and 0.496956381; the answer is therefore 0.497 (the true value is approximately 0.496953563).

22. $\dfrac{x - \sin x}{x^3 \cos x} = \dfrac{x^3/3! - x^5/5! + x^7/7! - \cdots}{x^3 - x^5/2! + x^7/4! - \cdots} = \dfrac{1/3! - x^2/5! + \cdots}{1 - x^2/2! + \cdots} \to \dfrac{1}{6}$ as $x \to 0$.

Chapter 11 Miscellaneous

1. $a_n = \dfrac{1 + \left(1/n^2\right)}{1 + \left(4/n^2\right)}.$

4. $a_n = 0$ for all n, so the limit is zero.

7. $-1 \leq \sin x \leq 1$ for all x, so $\{a_n\} \to 0$ by the squeeze theorem.

10. Apply l'Hôpital's rule thrice to obtain the limit zero.

13. Because e^n/n increases without bound while e^{-n}/n approaches zero, the limit does not exist (it is also correct to say that the given expression approaches $+\infty$).

16. The ratio test involves finding the limit of

$$\left(\frac{n}{n+1}\right)^n (n+1)\left(n^2 + 2n\right)\left(n^2 + 2n - 1\right) \cdots \left(n^2 + 1\right)$$

as n increases without bound. The first factor approaches e, so the limit is $+\infty$. Therefore the series diverges.

22. The given series diverges by the nth term test: $\lim\limits_{n \to \infty} -\dfrac{2}{n^2} = 0$, and therefore $\lim\limits_{n \to \infty} 2^{-\left(2/n^2\right)} = 1 \neq 0$.

28. $\lim\limits_{x \to \infty} x \sin \dfrac{1}{x} = \lim\limits_{u \to 0+} \dfrac{\sin u}{u} = 1$, so the given series diverges by the nth term test.

31. Is this not the Maclaurin series for $f(x) = e^{2x}$?

34. Limit: $\frac{1}{4}|2x - 3|$. Interval of convergence: $(-\frac{1}{2}, \frac{7}{2})$.

40. Limit: $|x - 1|$. Interval of convergence: $(0, 2)$.

43. Use the ratio test.

46. If $\sum a_n$ converges, then $\{a_n\} \to 0$. So $a_n \leq 1$ for all $n \geq N$ (where N is some sufficiently large integer). So the series $\sum a_n$ (eventually) dominates $\sum (a_n)^2$; therefore the latter series also converges.

52. $\ln(1 + x) = x - \frac{1}{2}x^2 + \frac{1}{3}x^3 - \frac{1}{4}x^4 + \dots$. For three-place accuracy, we need the first five terms of this series, and we find that

$$0.182266666 < \ln(1.2) < 0.182330667.$$

So $\ln(1.2) = 0.182$ to three places.

55. $\displaystyle\int_0^1 \left(1 - \frac{1}{2}x + \frac{1}{6}x^2 - \frac{1}{24}x^3 + \frac{1}{120}x^4 - \frac{1}{720}x^5 + \dots\right) dx$

$\displaystyle = \left[x - \frac{1}{4}x^2 + \frac{1}{18}x^3 - \frac{1}{96}x^4 + \frac{1}{600}x^5 - \frac{1}{4320}x^6 + \dots\right]_0^1$

$\displaystyle 1 - \frac{1}{2!\,2} + \frac{1}{3!\,3} - \frac{1}{4!\,4} + \frac{1}{5!\,5} - \frac{1}{6!\,6} + \dots \approx 0.796599599297.$

58. $\displaystyle\int_0^x \frac{dt}{1 - t^2} = \int_0^x \left(1 + t^2 + t^4 + t^6 + \dots\right) dt$

$\displaystyle = \left[t + \frac{1}{3}t^3 + \frac{1}{5}t^5 + \frac{1}{7}t^7 + \dots\right]_0^x$

$\displaystyle = x + \frac{1}{3}x^3 + \frac{1}{5}x^5 + \frac{1}{7}x^7 + \dots = \sum_{n=0}^{\infty} \frac{x^{2n+1}}{2n+1}.$

64. $\displaystyle P_k = \prod_{n=2}^{k} \frac{n^2}{n^2 - 1} = \left(\frac{2 \cdot 2}{1 \cdot 3}\right)\left(\frac{3 \cdot 3}{2 \cdot 4}\right)\left(\frac{4 \cdot 4}{3 \cdot 5}\right) \dots \left(\frac{k \cdot k}{(k-1) \cdot (k+1)}\right) = \frac{2k}{k+1} \to 2$ as $k \to \infty$.

So $\displaystyle\prod_{n=2}^{\infty} \frac{n^2}{n^2 - 1} = 2$.

1.   Solve the first equation for $t = x - 1$; substitute into the second: $y = 2(x - 1) - 1 = 2x - 3$.
Answer: $y = 2x - 3$.

4.   First, $x^2 = t$, so $y = 3x^2 - 2$, $x \geq 0$.

7.   For $x = e^t$ we may conclude that $y = 4e^{2t} = 4x^2$. So the answer is $y = 4x^2$, $x > 0$.

10.   Given: $x = 2\cosh t$, $y = 3\sinh t$. Then $(x/2)^2 = 1 + (y/3)^2$, which we may write in the standard form $\left(\dfrac{x}{2}\right)^2 - \left(\dfrac{y}{3}\right)^2 = 1$. But $x \geq 2$, so the graph is the right branch only of the hyperbola with center at $(0,0)$ and with vertices at $(-2,0)$ and $(2,0)$.

13.   First, $\dfrac{dy}{dx} = \dfrac{dy/dt}{dx/dt} = \dfrac{9t^2}{4t} = \dfrac{9t}{4}$. So when $t = 1$, the slope is $9/4$ at the point $P(3,5)$ of tangency. Thus an equation of the tangent line is $9x - 4y = 7$. The second derivative is $\dfrac{9}{16t}$, so the curve is concave upward at the point where $t = 1$.

16. $dy/dx = -e^{-2t}$; the point of tangency is $(1,1)$ and the slope there is -1. An equation of the tangent line is $x + y = 2$. The second derivative is $2e^{-3t}$, so the graph is concave upward everywhere.

19.   Here, $dr/d\theta = \sqrt{3}e^{\theta\sqrt{3}}$. Therefore $\cot \psi = \dfrac{1}{r} \cdot \dfrac{dr}{d\theta} = \left(e^{-\theta\sqrt{3}}\right)\left(\sqrt{3}\right)\left(e^{\theta\sqrt{3}}\right) = \sqrt{3}$. Therefore $\tan \psi = 1/\sqrt{3}$, and therefore $\psi = \pi/6$, a constant, independent of the value of θ.

22.   First, $y^2 = t^6 - 6t^4 + 9t^2$
$$= x^3 - 6x^2 + 9x$$
$$= x(x-3)^2, \qquad 0 \leq x \leq 4.$$

Now symmetry about the x-axis is obvious, as are the two intercepts. By implicit differentiation (we find both dy/dx and dx/dy) we may gain even more information about the graph, which is shown at the right.

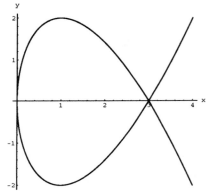

25.   If the slope at $P(x,y)$ is m, then $2y\dfrac{dy}{dx} = 4p$, so $y = \dfrac{2p}{dy/dx} = \dfrac{2p}{m}$. Therefore

$$4px = y^2 = \dfrac{4p^2}{m^2},$$

and so

$$x = \dfrac{p}{m^2}, \quad y = \dfrac{2p}{m}.$$

28. $|OC| = a - b$. So C has coordinates $x = (a-b)\cos t$, $y = (a-b)\sin t$. The arc length from the point of tangency to $A(a,0)$ is the same as that to P; denote it by s. Note that $s = ta$. Let α be the angle OCP and θ the angle supplementary to α, so that $\theta = \pi - \alpha$. Then $s = b\theta$. So $ta = b\theta$. The radius b is at the angle $-(\theta - t) = t - \theta$ from the horizontal, so P has coordinates

$$x = (a-b)\cos t + b\cos(t-\theta), \quad y = (a-b)\sin t + b\sin(t-\theta).$$

Now $\theta = \dfrac{a}{b}t$, so $t - \theta = t - \dfrac{a}{b}t = \dfrac{b-a}{b}t$. Hence P has the coordinates given in the statement of the problem.

1. Note first that $y = 2t^2 + 1$ is always positive, so the curve lies entirely above the x-axis. Moreover, as t goes from -1 to 1, $dx = 3t^2\, dt$ is positive, so the area is

$$A = \int_{-1}^{1} (2t^2 + 1)\,(3t^2)\ dt = \tfrac{22}{5}.$$

4. $A = \int_{1}^{0} -3e^{2t}\ dt = \tfrac{3}{2}\left(e^2 - 1\right).$

7. $V = \int_{-1}^{1} \pi\,(2t^2 + 1)^2\,(3t^2)\ dt.$

10. $V = \int_{\pi}^{0} -\pi e^{2t}\,\sin t\ dt = \dfrac{\pi}{5}\left(e^{2\pi} + 1\right).$

13. $\left(\dfrac{dx}{dt}\right)^2 + \left(\dfrac{dy}{dt}\right)^2 = 2,$ so $L = \int_{\pi/4}^{\pi/2} \sqrt{2}\ dt.$

16. $L = \int_{2\pi}^{4\pi} \sqrt{1 + \theta^2}\ d\theta$

$\quad = 2\pi\sqrt{1 + 16\pi^2} + \tfrac{1}{2}\ln\left(4\pi + \sqrt{1 + 16\pi^2}\right) - \pi\sqrt{1 + 4\pi^2} - \tfrac{1}{2}\ln\left(2\pi + \sqrt{1 + 4\pi^2}\right),$

$\quad$ which is approximately equal to 59.563022.

19. $A = 2\int_{0}^{1} 2\pi t^3 \sqrt{9t^4 + 4}\ dt.$

22. $A = \int_{0}^{\pi/2} 2\pi\,(e^{\theta}\cos\theta)\left(\sqrt{2}\,e^{\theta}\right)\ dt = \tfrac{2}{5}\pi\sqrt{2}\,(e^{\pi} - 2).$

25. (a) $A = 2\int_{0}^{\pi} (b\sin t)\,(a\sin t)\ dt.$ (b) $V = 2\int_{0}^{\pi} (b^2\sin^2 t)\,(a\sin t)\ dt.$

28. $A = \int_{0}^{2\pi} 2\pi a\,(b + a\cos t)\ dt = 4\pi^2 ab.$

31. Here, $ds = 3a\,|\sin t\,\cos t|\ dt$, so $A = (4\pi)\,(3a)\int_{0}^{\pi/2} (\sin t\,\cos t)\,(a\sin^3 t)\ dt.$

34. $ds = (t^2 + 3)\ dt$, so the arc length is $L = \int_{-3}^{3} (t^2 + 3)\ dt = 36.$

Section 12.3

4. $2\sqrt{65}$, $12\sqrt{5}$, $20\sqrt{2}$, $\langle 4, -8\rangle$, and $\mathbf{a} \cdot \mathbf{b} = -180$, so $\mathbf{a}$ and $\mathbf{b}$ are not perpendicular to each other.

10. $|\mathbf{a}| = 13$; choose $\mathbf{u} = \tfrac{5}{13}\mathbf{i} - \tfrac{12}{13}\mathbf{j}$; $\mathbf{v} = -\tfrac{5}{13}\mathbf{i} + \tfrac{12}{13}\mathbf{j}.$

13. $\langle 3, -2\rangle - \langle 3, 2\rangle = \langle 0, -4\rangle = -4\mathbf{j}.$

16. $-5\mathbf{i}.$

19. $\mathbf{a} \cdot \mathbf{b} = (2)(8) + (-1)(4) = 12 \neq 0;$ not perpendicular.

22. (a) $12\mathbf{i} - 20\mathbf{j};$ (b) $\tfrac{3}{4}\mathbf{i} - \tfrac{5}{4}\mathbf{j}$

25. $(2c\mathbf{i} - 4\mathbf{j}) \cdot (3\mathbf{i} + c\mathbf{j}) = 0$ when $6c - 4c = 0$, thus when $c = 0$.

28. $(r + s)\,\langle a_1, a_2\rangle = \langle (r + s)\,a_1, (r + s)\,a_2\rangle$

$\quad\qquad\qquad = \langle ra_1 + sa_1, ra_2 + sa_2\rangle$

$\quad\qquad\qquad = \langle ra_1, ra_2\rangle + \langle sa_1, sa_2\rangle$

$\quad\qquad\qquad = r\langle a_1, a_2\rangle + s\langle a_1, a_2\rangle = r\mathbf{a} + s\mathbf{a}.$

31. $(r\mathbf{a}) \cdot \mathbf{b} = \langle ra_1, ra_2 \rangle \cdot \langle b_1, b_2 \rangle$
    $= ra_1 b_1 + ra_2 b_2$
    $= r(a_1 b_1 + a_2 b_2) = r(\mathbf{a} \cdot \mathbf{b}).$

34. Use $\mathbf{v}_a = \mathbf{v}_g - \mathbf{w} = \langle -500, 0 \rangle - 25\sqrt{2} \langle -1, -1 \rangle.$

Section 12.4

1. Note that $\mathbf{r}$ is a constant vector.

4. $\mathbf{r}'(t) = \langle -\sin t, \cos t \rangle$ and $\mathbf{r}''(t) = \langle -\cos t, -\sin t \rangle.$

    So $\mathbf{r}'\left(\frac{\pi}{4}\right) = \left\langle -\frac{\sqrt{2}}{2}, \frac{\sqrt{2}}{2} \right\rangle$ and $\mathbf{r}''\left(\frac{\pi}{4}\right) = \left\langle -\frac{\sqrt{2}}{2}, -\frac{\sqrt{2}}{2} \right\rangle.$

7. $\mathbf{r}'(t) = \mathbf{i} \sec t \tan t + \mathbf{j} \sec^2 t$; $\mathbf{r}''(t) = \mathbf{i}\left(\sec^3 t + \sec t \tan^2 t\right) + \mathbf{j}\left(2\sec^2 t \tan t\right).$
    So $\mathbf{r}'(0) = \mathbf{j}$ and $\mathbf{r}''(0) = \mathbf{i}.$

10. $\int_1^e \langle \frac{1}{t}, -1 \rangle \, dt = \left[\langle \ln t, -t \rangle \right]_1^e = \langle 1, 1 - e \rangle.$

13. $(3\mathbf{i}) \cdot (2\mathbf{i} - 5t\,\mathbf{j}) + (3t\,\mathbf{i} - \mathbf{j}) \cdot (-5\mathbf{j}) = 11.$

16. $\mathbf{v}(t) = \mathbf{C}_1 = \langle 0, -2 \rangle.$ $\mathbf{r}(t) = \langle c, -2t \rangle + \langle C_2, C_3 \rangle$; $\mathbf{r}(0) = \langle 2, 3 \rangle = \langle C + C_2, C_3 \rangle$: $\mathbf{r}(t) = \langle 2, 3 - 2t \rangle.$

22. Now $dy/dt = 0$ when $gt = v_0 \sin \alpha$, so y_{max} occurs when $t = \frac{1}{g} v_0 \sin \alpha$; therefore

$$y_{max} = \frac{1}{2g}\left(v_0 \sin \alpha\right)^2.$$

Also $y = 0$ when $t = 0$ and when $t = \frac{1}{g}\left(2v_0 \sin \alpha\right).$ So the range is given by

$$R = \frac{1}{g}\left(v_0\right)^2 \sin 2\alpha.$$

Given the data of the problem—$\alpha = \pi/3$ and $R = 5280$—we find that

$$v_0 = (32)\left(110\sqrt{3}\right)^{1/2} \approx 441.7 \ \text{(ft/s) and}$$
$$y_{max} = 1320\sqrt{3} \approx 2286.31 \ \text{(ft)};$$

that is, about 0.433 miles.

25. The maximum altitude y_{max} occurs when $dy/dx = 0$. To keep the notation simple, we write v for v_0. Then $dy/dx = 0$ when $\dfrac{-gt + v\sin \alpha}{v \cos \alpha} = 0$, so that $t = (v \sin \alpha)/g$. We use this value of t in Equation (2) to find that

$$y_{max} = \frac{1}{2g}\left(v_0 \sin \alpha\right)^2.$$

Finally, as before, $R = \dfrac{\left(v_0 \sin 2\alpha\right)^2}{2g}.$ Results:

(a) $y_{max} = 100,$ $\quad R = 400\sqrt{3} \approx 692.82 \ \text{(ft)};$

(b) $y_{max} = 200,$ $\quad R = 800;$

(c) $y_{max} = 300,$ $\quad R = 400\sqrt{3}.$

28. See the diagram at the right. Given:

$$\mathbf{a} = -9.8\,\mathbf{j}$$
$$\mathbf{v}(t) = (v_0 \cos \alpha)\,\mathbf{i} + (v_0 \sin \alpha - 9.8t)\,\mathbf{j}$$
$$\mathbf{r}(t) = (v_0 \cos \alpha)\,t\,\mathbf{i} + (v_0 t \sin \alpha - 4.9t^2)\,\mathbf{j}$$

We require that at at the time T of interception,

$$800 - 4.9T^2 = 400$$
$$(v_0 \cos \alpha)\,T = 800 \quad \text{and}$$
$$v_0 T \sin \alpha - 4.9T^2 = 400$$

Solve: $T = \frac{20}{7}\sqrt{10} \approx 9.035$, $\alpha = \pi/4$, and
$v_0 = 28\sqrt{10}\sqrt{2} \approx 125.22$ m/s.

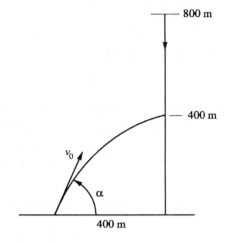

40. If $\mathbf{a}(t) = \langle 0,0 \rangle$ then $\mathbf{v}(t) = \mathbf{C} = \langle d_1, d_2 \rangle$. So

$$\mathbf{r}(t) = \langle c_1 t + c_3, c_2 t + c_4 \rangle.$$

The path of the particle is then parametrized by $x = c_1 t + c_3$, $y = c_2 t + c_4$. and is therefore a straight line. Moreover, the speed of the particle is $|\mathbf{v}(t)| = \sqrt{(c_1)^2 + (c_2)^2}$, a constant. This concludes the proof.

Section 12.5

4. $r = e^t$, $\theta = 2r$:
$$dr/dt = d^2r/dt^2 = e^t, \quad d\theta/dt = 2.$$
$$D_t \left(r^2 \frac{d\theta}{dt} \right) = D_t \left(2e^{2t} \right) = 4e^{2t}.$$
$$\mathbf{v}(t) = e^t \mathbf{u}_r + 2e^t \mathbf{u}_\theta; \quad \mathbf{a}(t) = -3e^t \mathbf{u}_r + 4e^t \mathbf{u}_\theta.$$

10. With the usual meanings of the symbols, we have $b^2 = a^2 \left(1 - e^2 \right)$. With the aid of a calculator and the result of Problem 8, we obtain these results: Its velocity at apogee is approximately 18.2007 miles per second; at perigee, it is approximately 18.8189 miles per second.

13. Equation (11), when applied to the moon, yields $(26.32)^2 = k \,(238,850)^3$.

For a satellite with period $T = \frac{1}{24}$ (of a day—one hour), it yields $1^2 = kr^3$ where r is the radius of the orbit of the satellite. Divide the second of these equations by the first and solve for $r \approx 3165$ miles, about 795 miles below the earth's surface. So it can't be done.

Chapter 12 Miscellaneous

4. Note that neither x nor y is ever negative.
So first we write

$$\sqrt{x} + \sqrt{y} = 1, \quad \text{then}$$
$$y = \left(1 - \sqrt{x} \right)^2, \quad 0 \le x \le 1.$$

The graph is shown at the right.

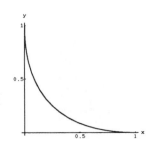

10. $dr/d\theta = \cos\theta$. By the formula in Equation (9) of Section 12.1,
$$\frac{dy}{dx} = \frac{\cos\theta + 2\sin\theta\,\cos\theta}{\cos^2\theta - \sin^2\theta - \sin\theta}.$$

At $\theta = \pi/3$, $dy/dx = -1$. Also
$$x = r\cos\theta = \tfrac14\left(2 + \sqrt3\right) \quad \text{and} \quad r\sin\theta = \tfrac14\left(3 + 2\sqrt3\right).$$

So an equation of the tangent line is $x + y = \tfrac14\left(5 + 3\sqrt3\right)$.

16. $dx/dt = -\tan t$, $dy/dt = 1$. So $ds = \sqrt{1 + \tan^2 t}\,dt = +\sec t\,dt$. Therefore
$$L = \int_0^{\pi/4} \sec t\,dt = \ln\left(1 + \sqrt2\right).$$

22. $dr/d\theta = 4\sin\theta$; $y = r\sin\theta$. $ds = \sqrt{r^2 + (dr/d\theta)^2}\,d\theta = 4\,d\theta$. Therefore the surface area is
$$A = \int_0^{\pi/2} (2\pi)\,(4\sin\theta\,\cos\theta)\,(4)\,d\theta = 16\pi.$$

28. From $r^2 = 2a^2\cos 2\theta$ we find that $2r\dfrac{dr}{d\theta} = -4a^2\sin 2\theta$, and thus that $\dfrac{dr}{d\theta} = -\dfrac1r\left(2a^2\sin 2\theta\right)$. So
$$\begin{aligned}
r^2 + \left(\frac{dr}{d\theta}\right)^2 &= r^2 + \frac{4a^4\sin^2 2\theta}{r^2}\\
&= \frac{r^4 + 4a^4\sin^2 2\theta}{r^2}\\
&= \frac{4a^4\cos^2 2\theta + 4a^4\sin^2 2\theta}{r^2} = \frac{4a^4}{r^2}.
\end{aligned}$$

Therefore $ds = \dfrac{2a^2}{|r|}\,d\theta$.
$$A = 2\int_{\theta=0}^{\pi/4} 2\pi y\,ds = 4\pi\int_0^{\pi/4}\frac{2a^2}{r}\,y\,d\theta = 4\pi\int_0^{\pi/4}\frac{2a^2}{r}\,r\sin\theta\,d\theta = 4\pi a^2\left(2 - \sqrt2\right).$$

34. First show that $a = -\omega^2 r$; conclude that
$$-\omega^2 x = x'' \quad \text{and} \quad -\omega^2 y = y'',$$
so that $x(t) = A\cos\omega t + B\sin\omega t$ and $y(t) = C\cos\omega t + D\sin\omega t$. Then use the given initial data to find A, B, C, and D.

37. We use primes to denote derivatives with respect to time t. From Equation (5) of Section 12.5,
$$\mathbf{a} = \left(r'' - (r)\,(\theta')^2\right)\mathbf{u}_r + \left(\frac1r\,(r^2\theta')'\right)\mathbf{u}_\theta.$$

Because $r^2\theta' = h$, $D_t\left(r^2\theta'\right) = 0$. So
$$\mathbf{a} = \left(r'' - (r)\,(\theta')^2\right)\mathbf{u}_r. \tag{$*$}$$

Therefore
$$\begin{aligned}
\frac{dr}{dt} &= \left(\frac{pe^2\sin\theta}{(1 + e\cos\theta)^2}\right)\left(\frac{d\theta}{dt}\right), \quad \text{so}\\
\frac{dr}{dt} &= \left(\frac{\sin\theta}{p}\right)\left(\frac{p^2 e^2}{(1 + e\cos\theta)^2}\right)\frac{d\theta}{dt}\\
&= \frac{\sin\theta}{p}\,r^2\frac{d\theta}{dt} = \frac{h}{p}\sin\theta,
\end{aligned}$$

and thus $\dfrac{d^2r}{dt^2} = \dfrac{h}{p}\left(\cos\theta\right)\dfrac{d\theta}{dt}$.

Now $1 + e\cos\theta = pe/r$, so $e\cos\theta = (pe - r)/r$; consequently $\cos\theta = (pe - r)/(er)$. Therefore

$$r'' = \left(\frac{h}{p}\right)\left(\frac{pe - r}{er}\right)\theta'.$$

Substitute this information in Equation $(*)$ and simplify; you should find that

$$\mathbf{a} = \frac{k}{r^2}\,\mathbf{u}_r$$

where k is the *constant* $-h^2/(pe)$. This is enough to establish the result of Problem 37.

4. $9\mathbf{i} - 3\mathbf{j} + 3\mathbf{k}$; $-14\mathbf{i} - 21\mathbf{j} + 43\mathbf{k}$; -34; $3\sqrt{21}$; $(2\mathbf{i} - 3\mathbf{j} + 5\mathbf{k})/\sqrt{38}$

7. $\dfrac{\mathbf{a} \cdot \mathbf{b}}{|\mathbf{a}|\,|\mathbf{b}|} = -\dfrac{13}{\sqrt{250}}$; $\theta = \cos^{-1}\left(-\dfrac{13}{50}\sqrt{10}\right)$—approximately 2.536 radians; that is, about $145.3°$.

10. $\cos\theta = -\frac{1}{10}\sqrt{2}$, so $\theta \approx 1.71269$; that is, $98°7'48''$.

16. $(x-3)^2 + (y-1)^2 + (z-2)^2 = 5^2$; that is, $x^2 - 6x + y^2 - 2y + z^2 - 4z - 11 = 0$.

19. If r is the radius, then $r^2 = 9 + 25 + 4 = 38$. So the equation of the sphere can be written in the form $(x-4)^2 + (y-5)^2 + (z+2)^2 = 38$.

22. $(x-4)^2 + \left(y-\frac{9}{2}\right)^2 + (z+5)^2 = \frac{85}{4}$. Center: $\left(4, \frac{9}{2}, -5\right)$. Radius: $\frac{1}{2}\sqrt{85}$.

28. $x^2 + y^2 + z^2 + 7 = 0$: No points satisfy this equation.

34. $\overrightarrow{PQ} = \langle 3, 1, 0\rangle$, $\overrightarrow{F} = \langle 1, 0, -1\rangle$: $W = \overrightarrow{F} \cdot \overrightarrow{PQ} = 3$.

46. We think of $\mathbf{a}$, $\mathbf{b}$, and $\mathbf{c}$ as having the same initial point. Now

$$\mathbf{c} = \frac{b\mathbf{a} + a\mathbf{b}}{a + b}$$

is a linear combination of $\mathbf{a}$ and $\mathbf{b}$, so $\mathbf{c}$ lies in the plane determined by $\mathbf{a}$ and $\mathbf{b}$. Let ϕ denote the angle between the vectors $\mathbf{a}$ and $\mathbf{c}$. Note that $\cos\phi = \dfrac{\mathbf{a} \cdot \mathbf{c}}{|\mathbf{a}|\,|\mathbf{c}|}$. Now

$$\mathbf{a} \cdot \mathbf{c} = \frac{\mathbf{a} \cdot (b\mathbf{a} + a\mathbf{b})}{a + b} = \frac{a^2 b + a^2 b \cos\theta}{a + b}$$

where θ is the angle between $\mathbf{a}$ and $\mathbf{b}$. So

$$\mathbf{a} \cdot \mathbf{c} = \frac{a^2 b}{a + b}(1 + \cos\theta) = \frac{2a^2 b}{a + b}\cos^2\left(\frac{\theta}{2}\right)$$

But $|\mathbf{a}| = a$; moreover,

$$\begin{aligned}
|\mathbf{c}| &= \frac{\sqrt{(b\mathbf{a} + a\mathbf{b}) \cdot (b\mathbf{a} + a\mathbf{b})}}{a + b} \\
&= \frac{\sqrt{b^2 a^2 + 2ab\,(\mathbf{a} \cdot \mathbf{b}) + a^2 b^2}}{a + b} \\
&= \frac{\sqrt{2a^2 b^2 + 2a^2 b^2 \cos\theta}}{a + b} \\
&= \frac{ab}{a + b}\sqrt{4\left(\frac{1 + \cos\theta}{2}\right)} = \frac{2ab}{a + b}\cos\left(\theta/2\right).
\end{aligned}$$

Therefore $\cos\phi = 2\dfrac{a^2 b}{a + b}\cos^2\left(\theta/2\right)\dfrac{a + b}{2a^2 b \cos\left(\theta/2\right)} = \cos\left(\theta/2\right)$. It now follows that $\phi = \theta/2$. Similarly, the angle between the vectors $\mathbf{b}$ and $\mathbf{c}$ is also equal to $\theta/2$. Consequently $\mathbf{c}$ bisects the angle between $\mathbf{a}$ and $\mathbf{b}$.

Alternative proof (using some geometry): Both $b\mathbf{a}$ and $a\mathbf{b}$ have length $|\mathbf{a}|\,|\mathbf{b}|$. So they form two sides of a rhombus. Their sum is a diagonal of the rhombus, and a theorem of geometry tells us that each diagonal of a rhombus bisects the angles at its ends.

52. $\cos^{-1}\left(\dfrac{1}{\sqrt{3}}\right)$ (about $54°44'8''$).

1. $\mathbf{a} \times \mathbf{b} = \begin{vmatrix} \mathbf{i} & \mathbf{j} & \mathbf{k} \\ 5 & -1 & -2 \\ -3 & 2 & 4 \end{vmatrix} = (-4+4)\,\mathbf{i} - (20-6)\,\mathbf{j} + (10-3)\,\mathbf{k}.$

4. $-24\mathbf{j} - 24\mathbf{k}$

7. You should find that $(\mathbf{a} \times \mathbf{b}) \times \mathbf{c} = -\mathbf{i} + \mathbf{j}$ while $\mathbf{a} \times (\mathbf{b} \times \mathbf{c}) = -\mathbf{k}.$

10. $\overrightarrow{PQ} = \langle 0, -1, 1 \rangle$ and $\overrightarrow{PR} = \langle -1, 0, 1 \rangle$. So the area A of the triangle PQR is given by

$$A = \tfrac{1}{2} |\overrightarrow{PQ} \times \overrightarrow{PR}| = \tfrac{1}{2}\,\text{ABS} \begin{vmatrix} \mathbf{i} & \mathbf{j} & \mathbf{k} \\ 0 & -1 & 1 \\ -1 & 0 & 1 \end{vmatrix} = \tfrac{1}{2} |\langle -1, -1, -1 \rangle| = \tfrac{1}{2}\sqrt{3}.$$

   (Because of the standard notation for determinants, we're using the programming language function $\text{ABS}(x) = |x|$ for clarity here.)

13. $V = \left| \overrightarrow{OP} \cdot \overrightarrow{OQ} \times \overrightarrow{OR} \right|$ where $\overrightarrow{OP} = \langle 1, 3, -2 \rangle$, $\overrightarrow{OQ} = \langle 2, 4, 5 \rangle$, and $\overrightarrow{OR} = \langle -3, -2, 2 \rangle$.

16. The side of length 255 is represented by the vector $\vec{r} = \langle 227.20666, 115.76758, 0 \rangle$ and the side of length 225 by the vector $\vec{s} = \langle 4.97678, 80.56982, 0 \rangle$. Now $\vec{r} \times \vec{s} = \langle 0, 0, 17729.85027 \rangle$, so the area of the triangle is

$$A = \tfrac{1}{2} |\vec{r} \times \vec{s}| = \tfrac{1}{2}(17729.85027) = 8864.925$$

   square feet (all these numbers are, of course, approximate).

22. $V = \dfrac{1}{6} \left| \overrightarrow{AP} \cdot \overrightarrow{AQ} \times \overrightarrow{AR} \right|$ and $V = \dfrac{d}{3} \cdot \dfrac{1}{2} \left| \overrightarrow{PQ} \times \overrightarrow{PR} \right|$. Therefore $d = \dfrac{\left| \overrightarrow{AP} \cdot \overrightarrow{AP} \times \overrightarrow{AR} \right|}{\left| \overrightarrow{PQ} \times \overrightarrow{PR} \right|}.$

   Here, $\overrightarrow{AP} = \langle 1, 3, 0 \rangle$, $\overrightarrow{AQ} = \langle 2, -1, 3 \rangle$, and $\overrightarrow{AR} = \langle -1, 0, 1 \rangle$. The value of the scalar triple product is -16, $\overrightarrow{PQ} \times \overrightarrow{PR}$ turns out to equal $\langle 5, -7, -11 \rangle$, and therefore $d = \frac{16}{195}\sqrt{195} \approx 1.14578.$

Section 13.3

4. A vector parallel to the line is $\overrightarrow{P_1 P_2} = \langle -6, 3, 5 \rangle$. If $\langle x, y, z \rangle$ is on the line through P_1 and P_2 then

$$\langle x, y, z \rangle = \overrightarrow{OP_1} + t\,\overrightarrow{P_1 P_2}$$

   where $O = (0, 0, 0)$ is the origin. That is,

$$\langle x, y, z \rangle - \langle 0, 0, 0 \rangle = t\,\langle -6, 3, 5 \rangle;$$

   it follows that parametric equations of the line in question are

$$x = -6t, \quad y = 3t, \quad z = 5t.$$

7. The equation of the plane has the form $x + 2y + 3z = D$ for some constant D. Because $(0, 0, 0)$ is in the plane, $D = 0$.

10. $y = 12$

13. A line perpendicular to the xy-plane is parallel to the unit vector $\mathbf{k}$. The x- and y-coordinates of all points on such a line are the same (in this case, $x = y = 1$) and the z-coordinates of such points are arbitrary. Parametric equations: $x = 1$, $y = 1$, $z = t$. Symmetric equations: $x - 1 = 0 = y - 1$.

16. The line contains $Q(3, 3, 1)$ (set $t = 1$) and $R(0, 2, 2)$ (set $t = 0$). Now $\overrightarrow{RQ} = \langle 3, 1, -1 \rangle$ and $\overrightarrow{OP} = \langle 2, -1, 5 \rangle$; therefore $\overrightarrow{OP} + t\,\overrightarrow{RQ} = \langle 2, -1, 5 \rangle + t\,\langle 3, 1, -1 \rangle$.

   Answers: $x = 2 + 3t$, $y = -1 + t$, $z = 5 - t$; $\quad \dfrac{x-2}{3} = \dfrac{y+1}{1} = \dfrac{z-5}{-1}.$

22. $x + y - 2z = -2$

25. A normal to the first plane is $\mathbf{i}$ and a normal to the second is $\mathbf{i} + \mathbf{j} + \mathbf{k}$. If θ is the angle between these two normals (by definition, this is the angle between the planes) then

$$\cos \theta = \frac{\langle 1, 0, 0 \rangle \cdot \langle 1, 1, 1 \rangle}{\sqrt{3}},$$

therefore $\theta = \arccos\left(1/\sqrt{3}\right)$ (about $54°44'8''$).

28. When $x = 0$, $y = \frac{1}{2}$, and $z = \frac{7}{2}$. Similarly, $(1, 1, 1)$ lies on the line. So one of the many correct answers is $x = 1 + t$, $y = 1 + \frac{t}{2}$, $z = 1 - \frac{5t}{2}$.

31. The xy-plane has equation $z = 0$, so the plane with equation $3x + 2y - z = 6$ meets it in the line with equation $3x + 2y = 6$. Two points on this line are $P(2, 0, 0)$ and $Q(0, 3, 0)$. In order to contain the third point $R(1, 1, 1)$, the plane in question must have normal vector $\overrightarrow{PQ} \times \overrightarrow{PR} = \langle 3, 2, 1 \rangle$. Hence it has an equation of the form $3x + 2y + z = D$ for some constant D. Because the point $(1, 1, 1)$ is in the plane, $D = 6$.

34. Suppose that the lines meet at (x, y, z). Then the simultaneous equations that result have solution $P(1, -1, 2)$ The first line also contains $Q(2, 1, 3)$ and the second also contains $R(3, 5, 6)$. A normal to the plane is $\overrightarrow{PQ} \times \overrightarrow{PR} = \langle 2, -2, 2 \rangle$; use $\langle 1, -1, 1 \rangle$ instead. Answer: $x - y + z = 4$.

40. You may assume without loss of generality that $a \neq 0$. Then the point $(-d_1/a, 0, 0)$ lies on the first plane. The desired result now follows.

Section 13.4

4. $\mathbf{v}(t) = 2t \langle 3, 4, -12 \rangle$; $\mathbf{a}(t) = 2\langle 3, 4, -12 \rangle$; $v(t) = |26t|$

10. $\mathbf{v}(t) = \langle t, 7 - t, 3t \rangle$; $\mathbf{r}(t) = \langle 5 + \frac{1}{2}t^2, 7t - \frac{1}{2}t^2, \frac{3}{2}t^2 \rangle$;

16. $\mathbf{v}(t) = \langle -1 + 3\cos 3t, -3\sin 3t, 4t - 7 \rangle$; $\mathbf{r}(t) = \langle 3 - t - \sin 3t, 3 + \cos 3t, 2t^2 - 7t \rangle$;

22. Let $\mathbf{v}(t)$ denote the velocity vector of the moving particle. If the speed $|\mathbf{v}(t)|$ of the particle is constant, then $\mathbf{v}(t) \cdot \mathbf{v}(t) = C$ where C is a constant. Now differentiate each side of this identity with respect to time.

25. Because $\mathbf{F} = k\mathbf{r}$, $\mathbf{r}$ and $\mathbf{a}$ are parallel. So

$$D_t (\mathbf{r} \times \mathbf{v}) = (\mathbf{r} \times \mathbf{a}) + (\mathbf{v} \times \mathbf{v}) = 0 + 0 = 0.$$

Therefore $\mathbf{r} \times \mathbf{v} = \mathbf{C}$, a constant vector. Consequently the vector $\mathbf{r}$ is always perpendicular to the constant vector $\mathbf{C}$. This holds for every point on the trajectory of the particle, and thus every point on the trajectory lies in the plane through the origin with normal vector $\mathbf{C}$.

28. Situate the gun at the origin, north the positive y-direction, east the positive x-direction, upward the positive z-direction. Assume that the gun is fired at time $t = 0$ (seconds) with angle α of elevation and lateral deviation θ measured counterclockwise from the positive y-axis. Note that θ will be rather close to zero. If T is the time of impact of a shell at the point $(0, 5000, 0)$, it is then easy to derive the equations

$$T = 6000 \cos \alpha \sin \theta,$$
$$T \cos \alpha \cos \theta = 10, \quad \text{and}$$
$$4T = 125 \sin \alpha.$$

To obtain a first approximation to a solution, assume that $\theta = 0$. The displayed equations then imply that

$$\sin 2\alpha = \tfrac{16}{25} \quad \text{and} \quad T = 10 \sec \alpha.$$

The first of these equations has two first-quadrant solutions:

$$\alpha \approx 19.89590975° \quad \text{and} \quad \alpha \approx 70.10409025°.$$

The corresponding values of T are

$$T \approx 10.63476324 \quad \text{and} \quad T \approx 29.38476324 \quad \text{(seconds).}$$

One may now continue in a very pragmatic way: Fire the gun due north with the smaller value of α. It's easy to show that the shell won't clear the hill. So the larger value of α must be used in any case. If the gun is fired due north with the larger value of α, the shell will strike the ground at

$$(x, y, z) \approx (71.95535922, 5000, 0).$$

So swivel the gun counterclockwise through an angle of

$$\theta = \arctan\left(\frac{71.95535922}{5000}\right) \approx 0.8244907643°$$

and *really* fire it this time. The results:

$$T = 29.38476324 \quad \text{and point of impact} \quad (x, y, z) \approx (0.0074499339, 4999.482323, 0.000001).$$

This is certainly close enough! You can also verify that the shell easily clears the hill unless the hill has an abnormal shape (the shell reaches a maximum altitude of more than 3450 feet).

A more sophisticated solution might proceed as follows. Obtain the approximate values of α and T using $\theta = 0$. Beginning with those values, iterate the following versions of the first equations (this is a method of repeated substitution):

$$T = \tfrac{125}{4}\sin\alpha,$$

$$\theta = \arcsin\left(\frac{T}{6000\cos\alpha}\right),$$

$$\alpha = \arccos\left(\frac{10}{T\cos\theta}\right).$$

A few iterations of these equations, in the order given, results in convergence to the values

$$T \approx 29.38430462,$$

$$\alpha \approx 70.10161954°, \quad \text{and}$$

$$\theta \approx 0.8244650319°.$$

The point of impact is $(-0.0000000078, 5000, 0.0000011)$ if these values are used. The errors in x and z are undoubtedly roundoff errors.

Section 13.5

1. $\displaystyle\int_0^\pi \sqrt{(6\cos 2t)^2 + (6\sin 2t)^2 + 64}\ dt = 10\pi.$

4. $(ds/dt)^2 = t^2 + (1/t)^2 + 2 = \left(t + t^{-1}\right)^2$; integrate ds from $t = 1$ to $t = 2$ to get $s = \frac{3}{2} + \ln 2.$

10. The curvature is $\displaystyle\frac{|(1)(2) - (0)(7)|}{(1^2 + 7^2)^{3/2}} = \frac{1}{250}\sqrt{2}.$

16. The curvature at x is $f(x) = \displaystyle\frac{2\,|x|^3}{(x^4 + 1)^{3/2}}.$ It suffices to consider the case in which $x > 0$; if so, then

$$f'(x) = \frac{6x^2\left(1 - x^4\right)}{(x^4 + 1)^{5/2}}.$$

It follows that $x = 1$ maximizes $f(x)$, its maximum value is $1/\sqrt{2}$, and the points on the graph of the equation $xy = 1$ where the curvature is maximal are $(1, 1)$ and $(-1, -1)$.

22. $\mathbf{a}(t) = \langle -3\pi^2 \sin \pi t, -3\pi^2 \cos \pi t \rangle$. So $a_T = 0$, $a_N = 3\pi^2$.

34. $\mathbf{v} = \langle 1, 2t, 3t^2 \rangle$; $\mathbf{a} = \langle 0, 2, 6t \rangle$. So $\mathbf{v} \times \mathbf{a} = \langle 6t^2, -6t, 2 \rangle$. $v = \sqrt{1 + 4t^2 + 9t^4}$;

$|\mathbf{v} \times \mathbf{a}| = \sqrt{36t^4 + 36t^2 + 4}$. Thus the curvature is $\dfrac{2\sqrt{9t^4 + 9t^2 + 1}}{(9t^4 + 4t^2 + 1)^{3/2}}$.

40. $|\mathbf{v} \times \mathbf{a}| = \sqrt{6}e^{2t}$, $v = |\mathbf{v}| = \sqrt{3}e^t$, and the curvature is $\frac{1}{3}\sqrt{2}e^{-t}$.

Therefore $\dfrac{dv}{dt} = a_T = \sqrt{3}e^t$ and $a_N = \sqrt{2}e^t$.

46. The curvature is $\dfrac{a\omega^2}{a^2\omega^2 + b^2}$. Also $\dfrac{dv}{dt} = 0$, so $a_T = 0$. Next, $a_N = v^2 = a\omega^2$. It follows that

$\mathbf{T}(t) = \dfrac{\langle -a\omega \sin \omega t, a\omega \cos \omega t, b \rangle}{\sqrt{a^2\omega^2 + b^2}}$; $\mathbf{N}(t) = -\langle \cos\omega t, \sin \omega t, 0 \rangle$; $a_T = 0$; $a_N = a\omega^2$.

49. $s = s(t) = \displaystyle\int_0^t ds = \int_0^t \sqrt{9 \sin^2 t + 9 \cos^2 t + 16}\, dt = 5t$. Therefore $t = s/5$ and the rest is easy.

Section 13.6

1. A plane with intercepts $x = \frac{20}{3}$, $y = 10$, $z = 2$.

4. A cylinder whose rulings are lines parallel to the z-axis. It meets the xy-plane in the hyperbola

$$x^2 - y^2 = 9.$$

The graph is shown at the right.

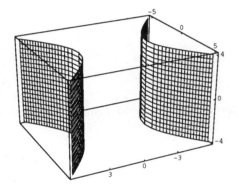

7. An elliptic paraboloid. The graph is shown at the right.

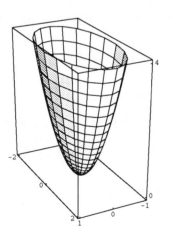

10. A circular cylinder with axis the x-axis.

16. Parabolic cylinder, parallel to the z-axis; its trace in the xz-plane is the parabola with vertex

$$x = 9 \quad \text{and} \quad z = 0.$$

The cylinder opens in the negative x-direction. The graph is shown at the right.

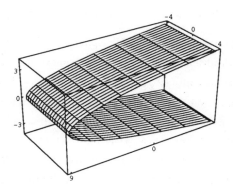

22. A cylinder having the graph

$$x = \sin y$$

as its trace in the xy-plane. The graph is shown at the right.

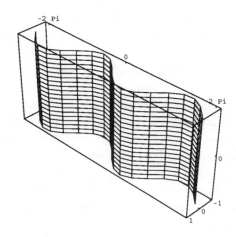

28. A hyperboloid of one sheet. Its graph is shown on the right.

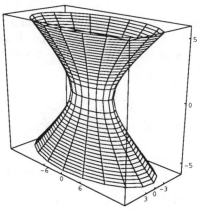

34. Replace x by $\sqrt{x^2 + y^2}$. Answer: A paraboloid opening along the negative z-axis and with equation

$$z = 4 - x^2 - y^2.$$

40. $4x^2 = y^2 + z^2$

46. Hyperbolas (both branches), except that the trace in the xy-plane itself consists of the coordinate axes.

52. Projection of the intersection: $2x^2 + 3z^2 = 5 - 3x^2 - 2z^2$. That is,

$$x^2 + z^2 = 1 : \quad \text{a circle.}$$

1. Cylindrical: $(0, \pi/2, 5)$; spherical: $(5, 0, \pi/2)$

4. Cylindrical: $(2\sqrt{2}, -\pi/4, 0)$; spherical: $(2\sqrt{2}, \pi/2, -\pi/4)$

10. Cylindrical: $(2\sqrt{5}, \tan^{-1}(-2), -12)$; spherical: $(2\sqrt{41}, \cos^{-1}(-\sqrt{41}/6), \tan^{-1}(-2))$

16. The lower nappe of the cone $z^2 = 3r^2$, vertex at the origin, axis the z-axis, opening downward.

22. From its Cartesian equation $z = 10 - 3x^2 - 3y^2$, we see that the surface is a circular paraboloid opening downward, with highest point $(0, 0, 10)$.

28. $r(\cos\theta + \sin\theta) = 4$; $\rho\sin\phi\cos\theta + \rho\sin\phi\sin\theta = 4$.

34. Take $\rho = 3960$ (miles). Atlanta: $\phi = 56.25°$, $\theta = -84.40°$.

    San Francisco: $\phi = 52.22°$, $\theta = -122.42°$.

    Therefore Atlanta is at $P_1(x_1, y_1, z_1)$ where $x_1 \approx 321.3$, $y_1 \approx -3276.9$, and $z_1 \approx 2200.1$;

    San Francisco is at $P_2(x_2, y_2, z_2)$ where $x_2 \approx -1677.8$, $y_1 \approx -2642.0$, and $z_2 \approx 2426.0$.

    The cosine of the central angle between the two cities is then $\dfrac{\left(\overrightarrow{OP}_1 \cdot \overrightarrow{OP}_2\right)}{\left|\overrightarrow{OP}_1\right|\left|\overrightarrow{OP}_2\right|} \approx 0.858073636$.

    This means that the angle itself is about $30.89903595°$; that is, about 0.5392897386 radians. The arc length between the two cities is the product of the radius of the earth and this angle (in radians), and that's approximately 2135.587365. Answer: The great circle distance between San Francisco and Atlanta is about 2136 miles.

37. We found the rectangular coordinates of the point of the journey nearest the North Pole to be approximately $(x, y, z) = (15.99750973, 26.41463504, 3959.879587)$. The central angle between this point and the North Pole has cosine z/R where $R = 3960$; that angle turns out to be about 0.00779839 radians. Multiply by R to get the closest approach to the North Pole in great circle terms, hardly different from the straight-line distance.

40. Maximize z given $ax + by + cz = 0$ and $x^2 + y^2 + z^2 = r^2$: $r = 3960$ is the radius of the earth and $ax + by + cz = 0$ is the equation of the plane through New York, London, and the center of the earth. We found $a = -8.93709 \times 10^6$, $b = 3.80960 \times 10^6$, and $c = 7.10888 \times 10^6$. The Lagrange multiplier method yields:
$$x = -\frac{acr}{\sqrt{a^2 + b^2}\sqrt{a^2 + b^2 + c^2}}, \quad y = -\frac{bcr}{\sqrt{a^2 + b^2}\sqrt{a^2 + b^2 + c^2}}, \quad \text{and} \quad z = \frac{r\sqrt{a^2 + b^2}}{\sqrt{a^2 + b^2 + c^2}}.$$
The latitude of this point is $53.806°$, somewhat farther north than London ($51.5°$).

Chapter 13 Miscellaneous

10. For $i = 1$ and $i = 2$, the line L_i passes through the point $P_i(x_i, y_i, z_i)$ and is parallel to the vector $\mathbf{v}_i = \langle a_i, b_i, c_i \rangle$. Let L be the line through P_1 parallel to $\mathbf{v}_2$. Now, by definition, the lines L_1 and L_2 are skew if and only they are not coplanar. But L_1 and L_2 are coplanar if and only if the plane determined by L_1 and L is the same as the plane determined by L_2 and L; that is, if and only if the plane determined by L_1 and L is the same as the plane that contains the segment $P_1 P_2$ and the line L. Thus L_1 and L_2 are coplanar if and only if $\overrightarrow{P_1P_2}$, $\mathbf{v}_1$, and $\mathbf{v}_2$ are coplanar, and this condition is equivalent (by a theorem from early in Chapter 13) to the condition
$$|\overrightarrow{P_1P_2} \cdot \mathbf{v}_1 \times \mathbf{v}_2| = 0.$$

That is, L_1 and L_2 are coplanar if and only if $\begin{vmatrix} x_1 - x_2 & y_1 - y_2 & z_1 - z_2 \\ a_1 & b_1 & c_1 \\ a_2 & b_2 & c_2 \end{vmatrix} = 0$. This establishes the conclusion given in Problem 10.

16. $x + 2y + 3z = 6$

22. $\dfrac{dx}{dt} = \dfrac{dr}{dt}\cos\theta - (r\sin\theta)\dfrac{d\theta}{dt}$ and $\dfrac{dy}{dt} = \dfrac{dr}{dt}\sin\theta + (r\cos\theta)\dfrac{d\theta}{dt}$. Now show that

$$\left(\frac{dx}{dt}\right)^2 + \left(\frac{dy}{dt}\right)^2 + \left(\frac{dz}{dt}\right)^2 = \left(\frac{dr}{dt}\right)^2 + \left(r\left(\frac{d\theta}{dt}\right)\right)^2 + \left(\frac{dz}{dt}\right)^2.$$

28. Replace y by r to obtain $(r-1)^2 + z^2 = 1$. Simplify this to $r^2 - 2r + z^2 = 0$, then convert to the rectangular equation $x^2 + y^2 + z^2 = 2\sqrt{x^2 + y^2}$.

34. Because $z = ax^2 + by^2$ is a paraboloid (not a hyperboloid), we know that $ab > 0$. The projection into the xy-plane of the intersection K has the equation $ax^2 - Ax + by^2 - By = 0$. We may also assume that $a > 0$ and $b > 0$ (multiply through by -1 if necessary). It follows by completing squares that this projection is either empty, a single point, or an ellipse. In the latter case, it follows from Problem 32 that K is an ellipse.

40. The curvature is

$$\kappa(t) = \frac{|x'(t)y''(t) - x''(t)y'(t)|}{\left[(x'(t))^2 + (y(t))^2\right]^{3/2}} = \frac{ab\sin^2 t + ab\cos^2 t}{\left(a^2\sin^2 t + b^2\cos^2 t\right)^{3/2}} = \frac{ab}{\left(a^2\sin^2 t + b^2\cos^2 t\right)^{3/2}}.$$

$$\kappa'(t) = \tfrac{3}{2}ab\left(a^2\sin^2 t + b^2\cos^2 t\right)^{-5/2}\left(2a^2\sin t\cos t - 2b^2\sin t\cos t\right) = \frac{3ab\left(\sin t\cos t\right)\left(a^2 - b^2\right)}{\left(a^2\sin^2 t + b^2\cos^2 t\right)^{5/2}}.$$

$\kappa'(t) = 0$ when $t = 0$, $\pi/2$, π, $3\pi/2$: $\kappa(0) = \dfrac{ab}{b^3} = \dfrac{a}{b^2} = \kappa(\pi)$, and $\kappa(\pi/2) = \dfrac{ab}{a^3} = \dfrac{b}{a^2} = \kappa(3\pi/2)$. Because $a > b$, κ is maximal when $t = 0$ and when $t = \pi$—at $(\pm a, 0)$— and minimal when $t = \pi/2$ and when $t = 3\pi/2$—at $(0, \pm b)$.

43. Because $y(x) = Ax + Bx^3 + Cx^5$ is an odd function, the condition $y(1) = 1$ will imply that $y(-1) = -1$. Because $y'(x) = A + 3Bx^2 + 5Cx^4$ is an even function, the condition $y'(1) = 0$ will imply that $y'(-1) = 0$ as well. Because the graph of y is symmetric about the origin, the condition that the curvature is zero at $(1,1)$ will imply that it is also zero at $(-1,-1)$. The curvature at (x, y) is

$$\frac{|6Bx + 20Cx^3|}{\left(1 + (A + 3Bx^2 + 5Cx^4)^2\right)^{3/2}}$$

by Problem 50 in Section 13.5, so the curvature at $(1,1)$ will be zero when $|6B + 20C| = 0$. Thus we obtain the simultaneous equations

$$\begin{array}{rcrcrcl} A &+& B &+& C &=& 1, \\ A &+& 3B &+& 5C &=& 0, \\ && 3B &+& 10C &=& 0. \end{array}$$

These are easy to solve for $A = \dfrac{15}{8}$, $B = -\dfrac{5}{4}$, and $C = \dfrac{3}{8}$. Thus the equation of the connecting curve is $y = \dfrac{15}{8}x - \dfrac{5}{4}x^3 + \dfrac{3}{8}x^5$.

Chapter 14: Partial Differentiation
Section 14.2

1. The domain is the coordinate plane.

4. The points in the plane on or within the circle with equation $x^2 + y^2 = 4$; that is, the points of the disk of radius 2 and centered at the origin.

7. All points in the plane except for those on either of the the two lines $y = x$ and $y = -x$.

10. All points in space for which $xyz > 0$. This consists of the interior of the first octant (the points where x, y, and z are all positive) together with three other octants (one where x is positive and y and z are both negative, etc.).

13. The plane that meets the xz-plane in the line $z = x$ and the yz-plane in the line $z = y$.

16. The paraboloid $z = 4 - r^2$ with highest point $(0, 0, 4)$, symmetric about the z-axis and opening downward.

19. The lower nappe of a $45°$ cone with vertex at $(0, 0, 10)$, symmetric about the z-axis, and with cylindrical equation $z = 10 - r$.

22. Some typical level curves of

$$f(x, y) = x^2 - y^2$$

are shown at the right.

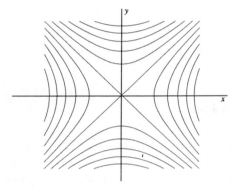

25. The level curves are vertical translates of the graph of $y = x^3$.

28. The level curves are circles centered at $(3, -2)$.

31. The level surfaces are circular paraboloids $z = C + r^2$ symmetric about the z-axis and opening upward. They are vertical translates of the "standard" parabolic surface with equation $z = r^2$.

34. The level surfaces are circular hyperboloids of two sheets symmetric about the z-axis, both nappes of the $45°$ cone $z = r$ with vertex at $(0, 0, 0)$ and symmetric about the z-axis, and circular hyperboloids of one sheet symmetric about the z-axis.

40. The traces of the surface in planes parallel to the yz-plane are vertical translates of the graph of the cubic equation $z = 2y^3 - 3y^2 - 12y$, which has one local maximum, one local minimum, and one inflection point. The traces of the surface in planes parallel to the xz-plane are vertical translates of the parabola $z = x^2$.

Section 14.3

4. 0

10. 0

16. 1, 1

28. On the line $x = 1$, $f(x, y)$ has limit 1. On the line $y = 1$, $f(x, y)$ has limit $\frac{1}{3}$. Therefore f has no limit at $(1, 1)$ and consequently f cannot be defined at $(1, 1)$ to be continuous there.

Section 14.4

4. $f_x(x,y) = (2x + x^2 y) e^{xy}, \quad f_y(x,y) = x^3 e^{xy}$

10. $f_x(x,y) = \dfrac{y}{1 + x^2 y^2}, \quad f_y(x,y) = \dfrac{x}{1 + x^2 y^2}$

16. $\dfrac{\partial f}{\partial u} = (4u - 4u^3 - 6uv^2) e^{-u^2 - v^2}, \quad \dfrac{\partial f}{\partial v} = (6v - 4u^2 v - 6v^3) e^{-u^2 - v^2}$

22. $z_{xy} = z_{yx} = 10x + 4y^3$

28. $z_{xy} = z_{yx} = (x^2 y + xy^2) \sec xy + 2(x + y) \sec xy \tan xy + 2(x^2 y + xy^2) \sec xy \tan^2 xy$

31. A normal to the plane at (x, y, z) is $\langle 2x, 2y, -1 \rangle$. So a normal to the plane at $(3, 4, 25)$ is $\langle 6, 8, -1 \rangle$. Therefore an equation of the tangent plane is $6x + 8y - z = 25$.

34. $2x + 2y - \pi z = 4 - \pi$

40. $3x - 4y - 5z = 0$

46. You should find that $u_{xx} = -m^2 e^{-(m^2 + n^2)kt} \sin mx \, \cos ny$.

49. $P_V = -\dfrac{nRT}{V^2}, \quad V_T = \dfrac{nR}{P}, \text{ and } T_P = \dfrac{V}{nR}$.

52. Let $P(a, b, c)$ be a point on the paraboloid $z = x^2 + y^2$. Note that $c = a^2 + b^2$, $z_x(P) = 2a$, and $z_y(P) = 2b$. Thus the tangent plane at P has equation

$$2a(x - a) + 2b(y - b) - (z - c) = 0;$$

substitution of $a^2 + b^2$ for c yields the equation

$$2(ax + by) = z + a^2 + b^2.$$

In the xy-plane, we have $z = 0$, so the tangent plane meets the xy-plane in the line with equation

$$2ax + 2by = a^2 + b^2.$$

Now $(x, y) = (a/2, b/2)$ satisfies this equation, so the point $Q(a/2, b/2)$ is on the line. The line has slope $-a/b$, so the normal to the line has slope b/a. The segment OQ has this slope, so OQ is a perpendicular from the origin O to this line. Its length is $\frac{1}{2}\sqrt{a^2 + b^2}$, so the line is tangent to the circle with center O and that same radius; that circle has equation $x^2 + y^2 = \frac{1}{4}(a^2 + b^2)$, and the result in the problem now follows.

55. (a) $\dfrac{\partial f_1}{\partial x} = \cos x \, \sinh(\pi - y), \qquad \dfrac{\partial^2 f_1}{\partial x^2} = -\sin x \, \sinh(\pi - y)$.

$\dfrac{\partial f_1}{\partial y} = -\sin x \, \cosh(\pi - y), \qquad \dfrac{\partial^2 f_1}{\partial y^2} = \sin x \, \sinh(\pi - y)$.

Clearly $\dfrac{\partial^2 f_1}{\partial x^2} + \dfrac{\partial^2 f_1}{\partial y^2} = 0$.

(b) $\dfrac{\partial^2 f_2}{\partial x^2} = 4 \sinh 2x \, \sin 2y$.

(c) $\dfrac{\partial^2 f_3}{\partial x^2} = -9 \sin 3x \, \sinh 3y$.

(d) $\dfrac{\partial^2 f_4}{\partial x^2} = -16 \sinh 4(\pi - x) \, \sin 4y$.

Section 14.5

4. $(-1, 0, -1)$

10. $(1, -1, -1)$.

13. The only possible extremum is $f(1,1) = 1$. Put $x = 1+h$, $y = 1+k$. Then $f(x,y) = h^2 + k^2 + 1$, so that $(1,1)$ yields 1 as the global minimum value of f.

16. Here, $f(x,y) \approx 3x^4 + 6y^4$ when either $|x|$ or $|y|$ is large, and therefore f has a global minimum but no global maximum. The critical points yield the values $f(0,0) = 7$, $f(0,1) = 9$, $f(-1,0) = 6$, and $f(-1,1) = 8$. Therefore $f(-1,0) = 7$ is the global minimum value of $z = f(x,y)$.

19. The only critical point of $f(x,y)$ occurs at $(1,-2)$. Because $f(x,y) \approx e^{-x^2-y^2}$ when either $|x|$ or $|y|$ is large and because $f(x,y) > 0$ for all x and y, f has no global minimum value, so in this case it has the global maximum value e^5.

22. Global maximum: $f(0,1) = 1 = f(1,1)$. Global minimum: $f\left(\frac{1}{2},0\right) = -\frac{1}{4}$.

28. We shall minimize $x^2 + y^2 + z^2$ given the side condition $xyz = 1$. Let $f(x,y) = x^2 + y^2 + \dfrac{1}{(xy)^2}$. The four critical points of f are $(-1,-1)$, $(-1,1)$, $(1,-1)$, and $(1,1)$. It's clear that all four yield minima. So there are four points on the surface $xyz = 1$ nearest the origin; they are $(1,-1,-1)$, $(-1,1,-1)$, $(1,1,1)$, and $(-1,-1,1)$.

31. Maximize $v = xyz$ given $x + 2y + 3z = 6$.

34. If the base of the box has dimensions x by y and the height of the box is z, then we must maximize the cost $C = 3xy + 4xz + 4yz$ subject to the condition $xyz = 48$. Solve this last equation for z and substitute to obtain

$$C = C(x,y) = 3xy + \frac{192}{x} + \frac{192}{y}, \qquad x > 0, \ y > 0.$$

Clearly $C \to \infty$ as $x \to 0^+$, as $y \to 0^+$, or as x or y increases without bound. Therefore C does have a global minimum value, and at a point where $C_x = 0 = C_y$. Now

$$C_x(x,y) = 3y - \frac{192}{x^2}, \quad C_y(x,y) = 3x - \frac{192}{y^2}.$$

Both partials vanish when $3x^2 y = 192 = 3xy^2$, which implies that $y = x$ (not $y = -x$), and therefore $x^3 = 64$. Thus the dimensions of the box of minimum cost are base 4 ft by 4 ft, height 3 ft.

37. If the box has length x and its other two dimensions are y and z, then its girth is $2y + 2z$. The post office condition then amounts to $x + 2y + 2z \leq 108$; of course, to get the largest possible box, we take $x + 2y + 2z = 108$. The volume V of the box takes the form

$$V = V(y,z) = (108 - 2y - 2z)\,yz,$$

and the condition that $V_y(y,z) = 0 = V_z(y,z)$ leads to the equation $(z - 54)^2 = (y - 54)^2$. The possibility that $z - 54 = -(y - 54)$ leads to an absurdity, so $y = z$, and the answer in the text follows.

40. Use the hemisphere $x^2 + y^2 + z^2 = R^2$, $z \geq 0$. We are to maximize $V = (2x)(2y)(z)$ given $x^2 + y^2 + z^2 = R^2$, with each of x, y, and z in the closed interval $[0, R]$.

$$V = V(x,y) = 4xy\sqrt{R^2 - x^2 - y^2};$$
$$V_x(x,y) = \frac{4R^2 y - 8x^2 y - 4y^3}{\sqrt{R^2 - x^2 - y^2}} \quad \text{and} \quad V_y(x,y) = \frac{4R^2 x - 8xy^2 - 4x^3}{\sqrt{R^2 - x^2 - y^2}}.$$

Both partials are zero when $4y\left(R^2 - 2x^2 - y^2\right) = 0$ and $4x\left(R^2 - 2y^2 - x^2\right) = 0$. Because neither $x = 0$ nor $y = 0$ can yield a maximum, we find that $2x^2 + y^2 = R^2 = x^2 + 2y^2$. It follows that $x = y$, and subsequently that $x = y = z = \frac{1}{3}R\sqrt{3}$. Answer: $V_{\max} = 4xyz = \frac{4}{9}R^3\sqrt{3}$.

46. See the diagram on the right. Given: $xyz = 12$. We minimize cost $C = 3xy + 2xz + 2yz$.

$$C(x,y) = 3xy + 2(x+y) \cdot \frac{12}{xy} = 3xy + \frac{24}{x} + \frac{24}{y}.$$

$$\frac{\partial C}{\partial x} = 3y - \frac{24}{x^2} \text{ and}$$

$$\frac{\partial C}{\partial y} = 3x - \frac{24}{y^2}.$$

Both partial derivatives vanish when $x = y = 2$, and there $z = 3$. The minimum cost is \$36.

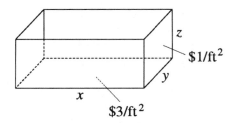

z

$\$1/\text{ft}^2$

y

x

$\$3/\text{ft}^2$

49. See the diagram at the right. We maximize the volume $V = xyz$ given $4x + 4y + 4z = 12$:

$$V(x,y) = xy(3 - x - y) = 3xy - x^2y - xy^2.$$

The critical points are $(0,0)$, $(3,0)$, $(0,3)$, and $(1,1)$. The maximum volume is 1, for a cubical box 1 m on a side.

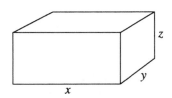

z

y

x

52. See the diagram at the right. We minimize the surface area $A = 2 \cdot 2\pi \frac{r}{2}\sqrt{r^2 + z^2} + 2\pi rh$ given fixed volume $V = \frac{2}{3}\pi r^2 z + \pi r^2 h$:

$$A(r,z) = 2\pi r \left(\frac{V}{\pi r^2} - \frac{2z}{3} + \sqrt{r^2 + z^2} \right).$$

When $\frac{\partial A}{\partial z} = 0$, $2r = z\sqrt{5}$ and $\sqrt{r^2 + z^2} = \frac{3}{2}z$. After considerable algebraic manipulation, we discover that when $\frac{\partial A}{\partial r} = 0$, $r^2 = \frac{5}{4}z^2$; since

$$z \neq 0, z = \left(\frac{12V}{25\pi} \right)^{1/3}, r = \frac{\sqrt{5}}{2} \left(\frac{12V}{25\pi} \right)^{1/3},$$

and

$$h = \left(\frac{20V}{9\pi} \right)^{1/3} - \left(\frac{24V}{225\pi} \right)^{1/3} \approx (0.56719)V^{1/3}.$$

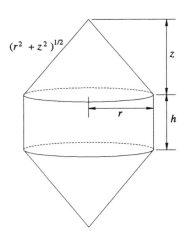

$(r^2 + z^2)^{1/2}$

z

r

h

At this point $z = \left(\frac{12V}{25\pi} \right)^{1/3} \approx (0.534602)V^{1/3}$ and $r = \frac{\sqrt{5}}{2}z \approx (0.597703)V^{1/3}$. The minimum value of the surface area there is (after more algebraic simplification) $A = \left(18\pi\sqrt{5}V^2 \right)^{1/3}$.

55. Minimize $f(x,y) = x^2 + (y-1)^2 + x^2 + y^2 + (x-2)^2 + y^2$
$$= 3x^2 - 4x + 3y^2 - 2y + 5.$$

$$\frac{\partial f}{\partial x} = 6x - 4, \qquad \frac{\partial f}{\partial y} = 6y - 2.$$

Both vanish at $\left(\frac{2}{3}, \frac{1}{3} \right)$. It is intuitively clear that f has a global minimum and no maximum, so the answer is $\left(\frac{2}{3}, \frac{1}{3} \right)$.

58. See the diagram at the right. We maximize the volume

$$V = xyz$$

$$\text{given } x^2 + y^2 + z^2 = L^2 :$$

$$V = V(x, y) = xy\sqrt{L^2 - x^2 - y^2}.$$

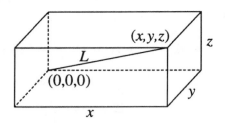

Both partial derivatives vanish when $x = y$ (since both x and y are positive), so by symmetry, $x = y = z$. Hence $x = L\sqrt{3}$, and the maximum volume is $3L^3\sqrt{3}$.

Section 14.6

4. $dw = (ye^{x+y} + xye^{x+y})\, dx + (xe^{x+y} + xye^{x+y})\, dy.$

10. $dw = ye^{uv}\, dx + xe^{uv}\, dy + xyve^{uv}\, du + xyue^{uv}\, dv.$

16. $dw = (1 - 2p^2)\, qre^{-(p^2+q^2+r^2)}\, dp + (1 - 2q^2)\, pre^{-(p^2+q^2+r^2)}\, dq + (1 - 2r^2)\, pqe^{-(p^2+q^2+r^2)}\, dr.$

19. $df = -\dfrac{dx + dy}{(1 + x + y)^2}$, so $\Delta f \approx -\dfrac{0.02 + 0.05}{(1 + 3 + 6)^2} = -0.0007$ (the true value is approximately -0.000695).

22. $df = \dfrac{(x + y + z)\, yz - xyz}{(x + y + z)^2}\, dx + \dfrac{(x + y + z)\, xz - xyz}{(x + y + z)^2}\, dy + \dfrac{(x + y + z)\, xy - xyz}{(x + y + z)^2}\, dz.$ Take $x = 2$, $y = 3$, $z = 5$, $dx = -0.02$, $dy = 0.03$, and $dz = -0.03$. Then $\Delta f \approx \frac{1}{10}(-0.24 + 0.21 - 0.09) = -0.012.$ (The true value is approximately -0.012323.)

28. Think of y as implicitly defined as a function of x by means of the equation $2x^3 + 2y^3 = 9xy$. It follows that

$$\frac{dy}{dx} = \frac{3y - 2x^2}{2y^2 - 3x}$$

and therefore that the slope of the line tangent to the graph of $2x^3 + 2y^3 = 9xy$ at the point $(1, 2)$ is $\frac{4}{5}$. The straight line through $(1, 2)$ with that slope has equation

$$y = 2 + \tfrac{4}{5}(x - 1);$$

when $x = 1.1$, we have $y_{\text{curve}} - y_{\text{line}} = 2 + \frac{4}{5}(0.1) = 2.08$. The true value of y is approximately 2.0757625.

34. -229.768 (cm^3)

Section 14.7

1. $\dfrac{dw}{dt} = \dfrac{dw}{dx} \cdot \dfrac{dx}{dt} + \dfrac{dw}{dy} \cdot \dfrac{dy}{dt} = -2xe^{-x^2-y^2} - yt^{-1/2}e^{-x^2-y^2} = -(2t + 1)\, e^{-t^2-t}.$

4. $\dfrac{dw}{dt} = \dfrac{2t}{1 + t^2}.$

10. $r_x = y^2 + z^2 + 2(xy + yz + xz) - 4x - 2y - 2z;$
    $r_y = x^2 + z^2 + 2(xy + yz + xz) - 2x - 4y - 2z;$
    $r_z = x^2 + y^2 + 2(xy + yz + xz) - 2x - 2y - 4z.$

16. $w_x = 6x$ and $w_y = 6y.$

22. $x - y - z = 0$

1. $\langle 3, -7 \rangle$

4. $\langle \frac{1}{8}\pi\sqrt{2}, -\frac{3}{8}\pi\sqrt{2} \rangle$

10. $\langle 160, -240, 400 \rangle$

16. $\frac{4}{3}\sqrt{3}$

22. The maximum value is $\dfrac{1}{\sqrt{5}}$; it occurs in the direction $\langle 2, 1 \rangle$.

28. $6x + 18y + 15z = 73$

40. Let $F(x, y, z) = z^2 - x^2 - y^2$ and $G(x, y, z) = 2x + 3y + 4z + 2$. First,

$$\nabla F \times \nabla G = \langle -8y - 6z, 4z + 8x, -6x + 4y \rangle.$$

For the tangent line to be horizontal, we require $-6x + 4y = 0$—that is, $3x = 2y$. So we have the three equations

$$3x = 2y,$$
$$z^2 = x^2 + y^2, \quad \text{and}$$
$$2x + 3y + 4z = -2.$$

Their simultaneous solution yields the answers:

$$\text{Low point:} \quad x = \frac{52 + 16\sqrt{13}}{39} \approx 2.812534,$$
$$y = \frac{26 + 8\sqrt{13}}{13} \approx 4.218801,$$
$$z = \frac{-8 - 2\sqrt{13}}{3} \approx -5.070368.$$
$$\text{High point:} \quad x = \frac{52 - 16\sqrt{13}}{39} \approx -0.145867,$$
$$y = \frac{26 - 8\sqrt{13}}{13} \approx -0.218801,$$
$$z = \frac{-8 + 2\sqrt{13}}{3} \approx -0.262966.$$

43. The hill is steepest in the direction of $\nabla z(-100, -100) = \langle 0.6, 0.8 \rangle$. The slope of the hill in that direction is $|\langle 0.6, 0.8 \rangle| = 1$, making your initial angle of climb $45°$. The compass heading of this direction is $\pi/2 - \arctan\frac{4}{3}$ radians, or about $36°52'12''$.

Section 14.9

1. The method leads to the simultaneous equations $2x = 2\lambda x$ and $2y + 2\lambda y = 0$. Their solutions: $x = 0$ or $\lambda = 1$.

4. The method yields $\langle 8x, 18y \rangle = \lambda\langle 2x, 2y \rangle$. It follows that one of x and y is zero. So the minimum occurs at $(-1, 0)$ and at $(1, 0)$, where the value of f is 4. The maximum value is 9, and occurs at $(0, 1)$ and at $(0, -1)$.

7. You should find that $2\lambda x = 8\lambda y = 18\lambda z = 1$. This leads to $z^2 = \frac{16}{49}$ while $x = 9z$ and $y = \frac{9}{4}z$.

10. We let $g(x, y, z) = x^2 + y^2 - 1$ and $h(x, y, z) = 2x + 2y + z - 5$. The equation $\nabla f = \lambda\nabla g + \mu\nabla h$ then takes the form

$$\langle 0, 0, 1 \rangle = \lambda\langle 2x, 2y, 0 \rangle + \mu\langle 2, 2, 1 \rangle.$$

It follows that $\lambda x = -1 = \lambda y$; because $\lambda \neq 0$, $x = y$. So $x = y = \frac{1}{2}\sqrt{2}$ or $x = y = -\frac{1}{2}\sqrt{2}$. In the first case we see that $z = 5 - 2\sqrt{2}$; in the second, $z = 5 + 2\sqrt{2}$. It follows that the maximum is $5 + 2\sqrt{2}$ and that the minimum is $5 - 2\sqrt{2}$.

16. (Problem 32 of Section 14.5) Frontage: x. Depth: y. Height: z. Given $xyz = 8000$, minimize the cost $C(x, y, z) = 2xy + 4xz + 8yz$. The method yields

$$\langle 2y + 4z, 2x + 8z, 4x + 8y \rangle = \lambda \langle yz, xz, xy \rangle.$$

Because x, y, and z are positive, we find that

$$\lambda xyz = 2xy + 4xz = 2xy + 8yz = 4xz + 8yz$$

and thus that $x = 2y = 4z$. Thus $8z^3 = 8000$, and so we find that $z = 10$, $y = 20$, and $x = 40$. Answer: The building should have a frontage of 40 feet, depth 20 feet, and height 10 feet. (This is a minimum by the usual argument—if any of the variables is near zero, then one of the others is very large, and this produces a large value for the cost C.)

19. (Problem 40 of Section 14.5) We are to minimize volume $V = 4xyz$ subject to the constraint $g(x, y, z) = x^2 + y^2 + z^2 - R^2 = 0$:

$$\langle 4yz, 4xz, 4xy \rangle = \lambda \langle 2x, 2y, 2z \rangle.$$

Because x, y, and z are positive at the maximum, we quickly find that $x = y = z$. Then we use the condition $g(x, y, z) = 0$ to find the answer: $V_{\max}$ occurs when $x = y = z = R/3$ and $V_{\max} = \frac{4}{9}\sqrt{3}\,R^3$.

22. Maximize A by maximizing

$$f(x, y, z) = A^2 = s(s - x)(s - y)(s - z)$$

given the constraint $g(x, y, z) = 2s - x - y - z = 0$ (where s is constant). The method yields

$$\langle s(s - y)(s - z), s(s - x)(s - z), s(s - x)(s - y) \rangle = \lambda \langle 1, 1, 1 \rangle.$$

Therefore

$$s(s - x)(s - y) = s(s - x)(s - z) = s(s - y)(s - z).$$

Now $s \neq 0$, so

$$s^2 - sx - sy + xy = s^2 - sy - sz + yz = s^2 - sx - sz + xz;$$

it follows that

$$x(s - z) = y(s - z) \quad \text{and} \quad y(s - x) = z(s - x).$$

If $s = x$ (say), the triangle degenerates to a line segment:

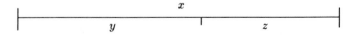

This yields minimum area zero. So at maximum, all three of $s - x$, $s - y$, and $s - z$ are positive. It follows that $x = y = z$, and therefore that the triangle with fixed perimeter and maximum area is an equilateral triangle.

28. Answers: The highest point is at $x = \dfrac{-15 - 9\sqrt{5}}{20} \approx -1.75623059$,

$$y = \dfrac{-15 - 9\sqrt{5}}{10} \approx -3.51248118,$$

$$z = \dfrac{9 + 3\sqrt{5}}{4} \approx 3.92705098.$$

The lowest point is at $x = \dfrac{-15 + 9\sqrt{5}}{20} \approx 0.256230590$,

$$y = \frac{-15 + 9\sqrt{5}}{10} \approx 0.512461179,$$

$$z = \frac{9 - 3\sqrt{5}}{4} \approx 0.572949017.$$

Section 14.10

1. $f(x, y) = 2(x + 1)^2 + (y - 2)^2 - 1$.

4. $y + 3 = 0 = x - 2$ at $(2, -3)$. $AC - B^2 = -1 < 0$, so f has no extrema.

7. $x^2 + y = 0 = x + y^2$ at $(0, 0)$ and at $(-1, -1)$. At the first critical point, $AC - B^2 = -9 < 0$; no extremum there. At $(-1, -1)$, $AC - B^2 = 27$ and $A = -6$. Therefore f has the local maximum value 4 at $(-1, -1)$.

10.   No extremum at $(0, 0)$; local maximum at $(1, 1)$.

16.   The equations $6x + 12y - 6 = 0 = 12x + 6y^2 + 6$ lead to $x = 1 - 2y$ and $y^2 - 4y + 3 = 0$, and thus $(-1, 1)$ and $(-5, 3)$ are the only two critical points. But $\Delta = 72y - 144$, so the first is not an extremum and the second yields a local minimum for f.

22.   Local (indeed, global) minimum at $(0, 0)$; no extremum at either of the other two critical points $(0, 1)$ and $(0, -1)$.

28.

(x, y)	A	B	C	Δ	Classification	z
$(0, 0)$	$2e$	0	$4e$	$8e^2$	Local minimum	0
$(0, 1)$	-2	0	-8	16	Local maximum	2
$(0, -1)$	-2	0	-8	16	Local maximum	2
$(1, 0)$	-4	0	2	-8	Saddle point	1
$(-1, 0)$	-4	0	2	-8	Saddle point	1

The following diagram gives a geometric summary of the critical points of the function.

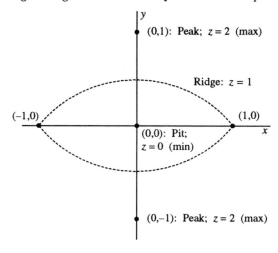

34.   At the point with polar coordinates (r, θ), f takes on the value

$$(r^2 \sin\theta \, \cos\theta)(\cos^2\theta - \sin^2\theta) = \tfrac{1}{2} r^2 \sin 2\theta \, \cos 2\theta = \tfrac{1}{4} r^2 \sin 4\theta.$$

As you walk around the z-axis on this surface, the sine function gives you four high spots and four low spots, so the graph near the origin is another dog saddle.

1. Transform the problem into polar form. Then the limit is $\lim\limits_{r\to 0} r^2 \sin\theta \cos\theta = 0$.

4. $g_x(0,0) = \lim\limits_{h\to 0} \dfrac{g(h,0)-g(0,0)}{h} = \lim\limits_{h\to 0} \dfrac{1}{h}\left(\dfrac{(h)(0)}{h^2+0^2}-0\right) = 0$.

7. The paraboloid is a level surface of $f(x,y,z) = x^2 + y^2 - z$ with gradient $\langle 2a, 2b, -1\rangle$ at the point (a,b,a^2+b^2). The normal line through that point thus has parametric equations
$$x = 2at + a,\ y = 2bt + b,\ z = a^2 + b^2 - t.$$
Set $x = 0 = y$ and $z = 1$, solve the last equation for t, and substitute in the other two to find
$$2a\left(a^2+b^2\right) - a = 0 = 2b\left(a^2+b^2\right) - b.$$
It then follows that $a^2 + b^2 = \frac{1}{2}$ or $a = 0 = b$.

10. You should obtain $u_{xx} = \dfrac{4\pi^2}{(4\pi kt)^{5/2}}\left(x^2 - 2kt\right)e^{-x^2/4kt}$.

16. First find the equation of the tangent plane. Let (a,b,c) be the point of tangency; note that $abc = 1$. Let $F(x,y,z) = xyz - 1$; then $\nabla F = \langle yz, xz, xy\rangle$, so that a normal to the plane is
$$\nabla F(a,b,c) = \langle bc, ac, ab\rangle,$$
and the plane therefore has equation $bc(x-a)+ac(y-b)+ab(z-c) = 0$; that is, $bcx+acy+abz = 3$. Its intercepts are $x_0 = \dfrac{3}{bc}$, $y_0 = \dfrac{3}{ac}$, and $z_0 = \dfrac{3}{ab}$. By the formula $V = \frac{1}{3}hB$, its volume is $\frac{9}{2}$.

22. Write $w = w(u,v) = \displaystyle\int_u^v f(t)\,dt$ where $u = g(x)$, $v = h(x)$. Then $F(x) = w(u(x), v(x))$. Consequently $F'(x) = w_u u_x + w_v v_x = -f(u)g'(x) + f(v)h'(x) = f(h(x))h'(x) - f(g(x))g'(x)$.

25. $z = 500 - (0.003)x^2 - (0.004)y^2$. $\nabla z = \langle -(0.006)x, -(0.008)y\rangle$. $\nabla z\big|_{(-100,-100)} = \langle\frac{3}{5}, \frac{4}{5}\rangle$. So in order to maintain a constant altitude, you should move in the direction of either $\langle -4, 3\rangle$ or $\langle 4, -3\rangle$.

28. $(0,b)$, $(0,-b)$, $(a,0)$, $(-a,0)$.

34. See the diagram below.

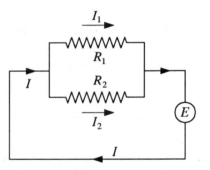

Let $x = I_1$ and $y = I_2$. Minimize $f(x,y) = R_1 x^2 + R_2 y^2$ given the constraint $g(x,y) = x + y - I = 0$ (I, R_1, and R_2 are all constants). The Lagrange multiplier method gives the vector equation $\langle 2R_1 x, 2R_2 y\rangle = \lambda\langle 1, 1\rangle$. It follows without any difficulty that
$$I_1 = x = \frac{R_2 I}{R_1 + R_2} \quad \text{and} \quad I_2 = y = \frac{R_1 I}{R_1 + R_2}.$$
The resistance R of the parallel circuit satisfies the equation $E = IR$, but $E = R_1 I_1 = R_2 I_2$, so
$$IR = \frac{R_1 R_2 I}{R_1 + R_2};$$
$$R = \frac{R_1 R_2}{R_2 + R_2};$$
$$\frac{1}{R} = \frac{R_1 + R_2}{R_1 R_2} = \frac{1}{R_1} + \frac{1}{R_2}.$$

40. We seek the extrema of $f(x,y) = x^2 y^2$ given the constraint $g(x,y) = x^2 + 4y^2 - 24 = 0$. The Lagrange multiplier method gives the following results: The maximum value 36 of f occurs at the four points $(2\sqrt{3}, \sqrt{3})$, $(-2\sqrt{3}, \sqrt{3})$, $(2\sqrt{3}, -\sqrt{3})$, and $(-2\sqrt{3}, -\sqrt{3})$. The minimum value 0 of f occurs at the four points $(2\sqrt{6}, 0)$, $(-2\sqrt{6}, 0)$, $(0, \sqrt{6})$, and $(0, -\sqrt{6})$.

46. Global minimum -2 at $(1, -2)$ and at $(-1, -2)$; saddle point at $(0, -2)$.

Chapter 15: Multiple Integrals

Section 15.1

1. $\displaystyle\int_0^2 \int_0^4 (3x + 4y)\, dx\, dy = \int_0^2 (24 + 16y)\, dy = 80.$

4. $\displaystyle\int_{-2}^1 \int_2^4 x^2 y^3\, dy\, dx = \int_{-2}^1 (64x^2 - 4x^2)\, dx = 180.$

10. $\displaystyle\int_0^{\pi/2} \cos x\, dx = 1.$

16. $\displaystyle\int_0^{\pi/2} (y - 1)\, dy = \tfrac{1}{8}\pi\,(\pi - 4).$

22. $\displaystyle\int_0^\pi \int_{-\pi/2}^{\pi/2} (\sin x\, \cos y)\, dy\, dx = \int_0^\pi 2 \sin x\, dx = 4.$

25. $\displaystyle\int_0^1 \int_0^1 x^n y^n\, dx\, dy = \int_0^1 \frac{1}{n+1} y^n\, dy = \frac{1}{(n+1)^2} \to 0 \text{ as } n \to +\infty.$

Section 15.2

1. $\displaystyle\int_0^1 \Big[y + xy \Big]_0^x dx = \int_0^1 (x + x^2)\, dx = \tfrac{5}{6}.$

4. $\displaystyle\int_0^2 \Big[\tfrac{1}{2}x^2 + xy \Big]_{y/2}^1 dy = \int_0^2 \big(\tfrac{1}{2} + y - \tfrac{5}{8}y^2 \big)\, dy = \tfrac{4}{3}.$

7. $\displaystyle\int_0^1 \int_x^{\sqrt{x}} (2x - y)\, dy\, dx = \int_0^1 \Big[2xy - \tfrac{1}{2}y^2 \Big]_x^{\sqrt{x}} dx = \int_0^1 \big(2x^{3/2} - \tfrac{1}{2}x - 2x^2 + \tfrac{1}{2}x^2 \big)\, dx$

$\displaystyle \qquad\qquad = \Big[\tfrac{4}{5}x^{5/2} - \tfrac{1}{4}x^2 - \tfrac{1}{2}x^3 \Big]_0^1 = \tfrac{1}{20}.$

10. 36

16. $\displaystyle\int_0^1 \int_y^{y^{1/4}} (x - 1)\, dx\, dy = -\tfrac{2}{15}.$

The region is shown at the right.

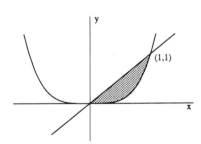

22. $\displaystyle\int_0^{\sqrt{\pi}} \int_0^x (\sin x^2)\, dy\, dx = 1.$

The region is shown at the right.

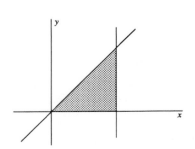

Section 15.3

1. The area is $A = \int_0^1 \int_{y^2}^y 1\, dx\, dy = \frac{1}{6}$.

4. $A = \int_1^3 \int_{2x+3}^{6x-x^2} 1\, dy\, dx = \frac{4}{3}$.

7. $A = \int_{-2}^2 \int_{2x^2-3}^{x^2+1} 1\, dy\, dx = \frac{32}{3}$.

10. $\pi - \frac{2}{3}$

13. The volume is $V = \int_0^1 \int_0^2 (y + e^x)\, dy\, dx = \int_0^1 \left[\frac{1}{2}y^2 + ye^x \right]_0^2 dx$

$$= \int_0^1 (2 + 2e^x)\, dx = \left[2x + 2e^x \right]_0^1 = 2 + 2e - 2 = 2e.$$

16. $V = \int_0^2 \int_0^{4-2y} (3x + 2y)\, dx\, dy = \int_0^2 \left[\frac{3}{2}x^2 + 2xy \right]_0^{4-2y} dy = \int_0^2 \left(24 - 16y + 2y^2 \right) dy = \frac{64}{3}$.

22. $\frac{837}{70}$

34. Region of integration: $x^2 + y^2 \leq 4$. Volume: 24π.

Section 15.4

1. $A = \int_0^{2\pi} \int_0^1 r\, dr\, d\theta$.

4. $A = \int_{-\pi/4}^{\pi/4} \int_0^{2\cos 2\theta} r\, dr\, d\theta = \pi/2$.

7. $A = 2 \int_{\pi/6}^{\pi/2} \int_0^{1-2\sin\theta} r\, dr\, d\theta$.

10. $\frac{3}{2}\pi$

13. $\int_0^{\pi/2} \int_0^1 \frac{r}{1+r^2}\, dr\, d\theta$

16. $\int_{\pi/4}^{\pi/2} \int_0^{\csc\theta} r^3 \cos^2\theta\, dr\, d\theta = \frac{1}{12}$.

19. The volume is $V = \int_0^{2\pi} \int_0^1 (2 + r\cos\theta + r\sin\theta)\, r\, dr\, d\theta = \int_0^{2\pi} \left[r^2 + \frac{1}{3}r^3 \cos\theta + \frac{1}{3}r^3 \sin\theta \right]_0^1 d\theta$

$$= \int_0^{2\pi} \left(1 + \frac{1}{3}\cos\theta + \frac{1}{3}\sin\theta \right) d\theta = \left[\theta + \frac{1}{3}\sin\theta - \frac{1}{3}\cos\theta \right]_0^{2\pi} = 2\pi.$$

22. $V = \int_0^{2\pi} \int_0^{1+\cos\theta} (1 + r\cos\theta)\, r\, dr\, d\theta = \frac{1}{6} \int_0^{2\pi} \left(3 + 9\cos^2\theta + 2\cos^4\theta \right) d\theta = \frac{11}{4}\pi$.

28. Domain of integration: the disk $x^2 + y^2 \leq 1$. $V = \int_0^{2\pi} \int_0^1 r\left(1 - r^2 \right) dr\, d\theta = \frac{1}{2}\pi$.

31. $V = \int_0^{\pi/2} \int_0^{\sqrt{2\sin 2\theta}} r^3\, dr\, d\theta$.

34. The polar form of the integral is $\int_0^{\pi/2} \int_0^\infty \frac{r}{(1+r^2)^2}\, dr\, d\theta$.

4. $M_y = \int_0^3 x\,(3 - x)\,dx = \frac{9}{2}.$

The area of the region is the same, so $\bar{x} = 1$. By symmetry, $\bar{y} = 1$ as well. Answer: $(1, 1)$.

10. By symmetry, $\bar{x} = 0$. $M_x = \frac{1}{2} \int_{-2}^2 (x^2 + 1)^2\,dx = \frac{206}{15}$. The area of the region is $\frac{28}{3}$, so $\bar{y} = \frac{103}{70}$.

16. $M = \int_0^1 \int_{x^2}^{\sqrt{x}} (x^2 + y^2)\,dy\,dx = \frac{6}{35}$, $\quad M_y = \int_0^1 \int_{x^2}^{\sqrt{x}} (x^3 + xy^2)\,dy\,dx = \frac{55}{504}$.

Hence $\bar{x} = \frac{275}{432}$. By symmetry, $\bar{y} = \bar{x}$.

22. We assume that $a > 0$. Then

$$M = \int_0^a \int_0^{a-x} (x^2 + y^2)\,dy\,dx = \frac{1}{6}a^4, \quad M_x = \int_0^a \int_0^{a-x} (x^2 y + y^3)\,dy\,dx = \frac{1}{15}a^5.$$

Because of the symmetry of x, y, and density in the region, it follows that $M_y = M_x$. Therefore $\bar{x} = \frac{2}{5}a = \bar{y}$.

28. $M = \frac{5}{3}\pi$, $M_x = 0$, $M_y = \frac{7}{4}\pi$, $\bar{x} = \frac{21}{20}$, $\bar{y} = 0$.

34. $\sqrt{3} + \frac{4}{3}\pi \approx 5.90841$.

40. $M = k\pi$ by inspection. With the aid of Formula 113 from the endpapers of the text, $I_y = \frac{5}{4}k\pi$ and $I_x = \frac{1}{4}k\pi$. Therefore $\hat{x} = \frac{1}{2}\sqrt{5}$ and $\hat{y} = \frac{1}{2}$.

43. Because $y = \sqrt{r^2 - x^2}$, $1 + \left(\dfrac{dy}{dx}\right)^2 = \dfrac{r^2}{r^2 - x^2}$. So $M_y = \int_0^r x\,\dfrac{r}{\sqrt{r^2 - x^2}}\,dx = r^2$. The length of the quarter-circle is $\frac{1}{2}\pi r$, so $\bar{x} = 2r/\pi$. By symmetry, $\bar{y} = \bar{x}$.

46. $V = \left(2\pi\dfrac{r}{3}\right)\left(\frac{1}{2}rh\right) = \frac{1}{3}\pi r^2 h.$

49. With the same notational change as in the solution to Problem 48, the radius of revolution is $\frac{1}{2}(r + s)$, so the lateral area is $A = 2\pi \left(\dfrac{r + s}{2}\right)\sqrt{(s - r)^2 + h^2} = \pi\,(r + s)\,L.$

52. (a) $A = \int_0^h \sqrt{2py}\,dy = \frac{2}{3}h^{3/2}\sqrt{2p}$. Now $r^2 = 2ph$, so $A = \frac{2}{3}h\sqrt{2ph} = \frac{2}{3}rh.$

$M_y = \int_0^h \frac{1}{2}(2py)\,dy = \frac{1}{2}ph^2$, so $\bar{x} = \dfrac{ph^2/2}{2rh/3} = \dfrac{3ph}{4r}$. But $ph = \frac{1}{2}r^2$, so $\bar{x} = \frac{3}{8}r.$

(b) $V = 2\pi x A = 2\pi M_y = \pi p h^2$. But $ph = \frac{1}{2}r^2$, so $V = \frac{1}{2}\pi r^2 h.$

Section 15.6

1. $\displaystyle\iiint_T f\,(x, y, z)\,dV = \int_0^1 \int_0^3 \int_0^2 (x + y + z)\,dx\,dy\,dz = \int_0^1 \int_0^3 \left[\frac{1}{2}x^2 + xy + xz\right]_0^2 dy\,dz$

$= \int_0^1 \int_0^3 (2 + 2x + 2y)\,dy\,dz = \int_0^1 \left[2y + y^2 + xyz\right]_0^3 dz$

$= \int_0^1 (15 + 6z)\,dz = \left[15z + 3z^2\right]_0^1 = 18.$

4. $\displaystyle\int_{-2}^6 \int_0^2 (4 + 4y + 4z)\,dy\,dz = \int_{-2}^6 (8 + 8z)\,dz = \left[8z + 4z^2\right]_{-2}^6 = 192.$

10. $\displaystyle\int_{-1}^1 \int_{-2}^2 \int_{y^2}^{8-y^2} z\,dz\,dy\,dx = \int_{-1}^1 \int_{-2}^2 (32 - 8y^2)\,dy\,dx = \int_{-1}^1 (128 - \frac{128}{3})\,dx = \frac{512}{3}.$

16. $\frac{128}{3}$

22. $M_{yz} = 0 = M_{xz}$; $\quad M_{xy} = \frac{1}{4}\pi R^4$.

28. $\frac{1}{2}\pi R^4 H$

34. $M = ka^5$, $M_{yz} = \frac{7}{12}ka^6$, $\bar{x} = \bar{y} = \bar{z} = \frac{7}{12}a$.

40. $\frac{16}{45}$

Section 15.7

1. $V = \int_0^{2\pi}\int_0^2\int_{r^2}^4 r\,dz\,dr\,d\theta = 2\pi\int_0^2\left(4r - r^3\right)dr = 2\pi\left[2r^2 - \frac{1}{4}r^4\right]_0^2 = 8\pi.$

4. Use the sphere $r^2 + z^2 \le a^2$, $0 \le \theta \le 2\pi$. We find its moment of inertia about the x-axis:

$$I_z = \int_0^{2\pi}\int_0^a\int_{-\sqrt{a^2-r^2}}^{\sqrt{a^2-r^2}} r^3\,dz\,dr\,d\theta = 2\pi\int_0^a 2r^3\sqrt{a^2-r^2}\,dr.$$

Let $r = a\sin u$. This substitution yields

$$I_z = 4\pi a^5\int_0^{\pi/2}\left(\cos^2 u - \cos^4 u\right)(\sin u)\,du = 4\pi a^5\left[\frac{1}{5}\cos^5 u - \frac{1}{3}\cos^3 u\right]_0^{\pi/2} = \frac{8}{15}\pi a^5.$$

The answer may also be written in the form $I_z = \frac{2}{5}Ma^5$ where M is the mass of the sphere.

7. Mass: $M = \int_0^{2\pi}\int_0^a\int_0^h rz\,dz\,dr\,d\theta = \pi\int_0^a rh^2\,dr = \left[\frac{1}{2}\pi r^2 h^2\right]_0^a = \frac{1}{2}\pi a^2 h^2.$

10. The diagram at the right shows the intersection of the xy-plane with the sphere and the cylinder. The plane and sphere intersect in the curve $r = 2$ and the cylindrical equation of the sphere is $r^2 + z^2 = 4$. The plane intersects the cylinder in the curve $r = 2\cos\theta$. Thus the volume V of their intersection is given by

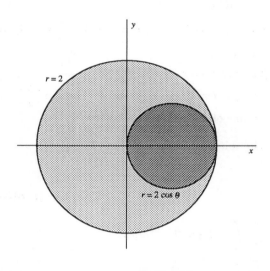

$$V = 4\int_0^{\pi/2}\int_0^{2\cos\theta}\int_0^{\sqrt{4-r^2}} r\,dz\,dr\,d\theta$$

$$= 4\int_0^{\pi/2}\int_0^{2\cos\theta} r\sqrt{4-r^2}\,dr\,d\theta$$

$$= 4\int_0^{\pi/2}\left[-\frac{1}{3}\left(4-r^2\right)^{3/2}\right]_0^{2\cos\theta}d\theta$$

$$= \frac{32}{3}\int_0^{\pi/2}\left(1 - \sin^3\theta\right)d\theta = \frac{16}{9}\left(3\pi - 4\right).$$

16. Let k denote the density of the cylinder and h its height. Then $m = \pi a^2 hk$. We choose coordinates so that the z-axis is the axis of symmetry of the cylinder and its base rests on the xy-plane. Then

$$I_z = \int_0^{2\pi}\int_0^a\int_0^h kr^3\,dz\,dr\,d\theta = \frac{1}{2}\pi a^4 hk = \left(\pi a^2 hk\right)\left(\frac{1}{2}a^2\right) = \frac{1}{2}ma^2.$$

22. Locate the hemisphere with its base in the xy-plane and symmetric about the z-axis. It follows that $M = \frac{1}{4}k\pi a^4$ and $M_{xy} = \frac{2}{15}\pi ka^5$; its centroid is at $\left(0, 0, \frac{8}{15}a\right)$.

25. $M_{xy} = \frac{1}{8}\pi a^4$

28. $I_z = \frac{1016}{21}\pi a^7$

Section 15.8

1. $\sqrt{1 + \left(z_x\right)^2 + \left(z_y\right)^2} = \sqrt{11}$. The area of the ellipse is 6π, so the answer is $6\pi\sqrt{11}$.

4. $\sqrt{1+(z_x)^2+(z_y)^2}=\sqrt{1+x^2}$. So the answer is

$$\int_0^1\int_0^x \sqrt{1+x^2}\,dy\,dx=\int_0^1 x\sqrt{1+x^2}\,dx=\left[\tfrac{1}{3}\left(1+x^2\right)^{3/2}\right]_0^1=\tfrac{1}{3}\left(2\sqrt{2}-1\right)\approx 0.609476.$$

10. $A=\tfrac{1}{6}\pi\left(17\sqrt{17}-1\right)\approx 36.1769$.

16. $A=2a^2\left(\pi-2\right)$, about 18% of the total surface area.

22. (b) The area in question is given by

$$A=\int_0^{2\pi}\int_0^{\pi/6} a^2\sin\phi\,d\phi\,d\theta=\left[2\pi-a^2\cos\phi\right]_0^{\pi/6}=2\pi a^2\left(1-\tfrac{1}{3}\sqrt{3}\right),$$

about 21.13% of the total surface area.

Section 15.9

4. $x=\tfrac{1}{2}\sqrt{u+v}$, $y=\tfrac{1}{2}\sqrt{u-v}$, and $J_T=\dfrac{\partial\left(x,y\right)}{\partial\left(u,v\right)}=-\dfrac{1}{8\sqrt{u^2-v^2}}$.

7. $J_T=\dfrac{\partial\left(x,y\right)}{\partial\left(u,v\right)}=-\tfrac{1}{5}$.

10. First, $x=\left(u^2v\right)^{-1/3}$, $y=\left(uv^2\right)^{-1/3}$, and $J_T=\dfrac{\partial\left(x,y\right)}{\partial\left(u,v\right)}=\dfrac{1}{3u^2v^2}$. It follows that the area of the region is $\tfrac{1}{8}$.

13. The Jacobian is $6r$.

16. $r=\sqrt{z},\ t=\sqrt{\dfrac{z}{x^2+y^2}}$, and $\theta=\arctan\left(\dfrac{y}{x}\right)$. The Jacobian of the transformation is $2\left(r/t\right)^3$. The volume of R turns out to be $45\pi/8$.

19. The spherical integral is $\displaystyle\lim_{a\to\infty}\int_0^{2\pi}\int_0^{\pi}\int_0^a \rho^3 e^{-k\rho^2}\sin\phi\,d\rho\,d\phi\,d\theta$.

Chapter 15 Miscellaneous

1. $\displaystyle\int_0^1\int_0^{x^3}\left(1+x^2\right)^{-1/2}dy\,dx$

4. $\displaystyle\int_0^4\int_0^{y^{3/2}} x\cos y^4\,dx\,dy$

7. $V=\displaystyle\int_0^1\int_u^{2-y}\left(x^2+y^2\right)dx\,dy$.

10. The domain of the integral will be the elliptical region bounded by the ellipse whose equation is $\left(x/2\right)^2+y^2=1$. Let us use the transformation $x=2r\cos\theta$, $y=r\sin\theta$. The Jacobian of this transformation is $2r$. The difference in the z-values of the two paraboloids turns out to be $8\left(1-r^2\right)$, and the volume of the region between them is 8π.

13. Use the same transformation as in the solution of Problem 10.

16. $M=\tfrac{6}{35}$; $M_y=\tfrac{55}{504}=M_x$; the centroid is at the point $\left(\tfrac{275}{432},\tfrac{275}{432}\right)\approx\left(0.636574,0.636574\right)$.

22. Note that $\bar{x}=\bar{y}$. The area of the quarter-ring is $A=\tfrac{1}{4}\left(\pi b^2-\pi a^2\right)=\tfrac{1}{4}\pi\left(b^2-a^2\right)$, and the volume generated by its rotation about the x-axis is $V=\tfrac{2}{3}\pi b^3-\tfrac{2}{3}\pi a^3$. Therefore

$$V=\tfrac{2}{3}\pi\left(b^3-a^3\right)=\left(2\pi\bar{y}\right)\frac{\pi}{4}\left(b^2-a^2\right).$$

(a) Consequently $\bar{y} = \dfrac{(2/3)\,\pi\,(b^3 - a^3)}{(1/2)\,\pi^2\,(b^2 - a^2)} = \dfrac{4\,(b^2 + ab + a^2)}{3\pi\,(b + a)} = \bar{x}.$

(b) $\displaystyle\lim_{b \to a} \bar{x} = \dfrac{12a^2}{(3\pi)\,(2a)} = \dfrac{2a}{\pi} = \lim_{b \to a} \bar{y}.$

25. $V = \displaystyle\int_0^{2\pi} \int_0^1 \left(r\sqrt{5 - r^2} - 2r^2\right)\, dr\, d\theta.$

28. $M = \frac{1}{48}a^6$

34. $2\pi^2 a^2 b$

46. $8\,(\pi - 2)$

52. The Jacobian of the transformation is 2; the integral becomes $\displaystyle\int_0^1 \int_{-v}^{v} 2e^{\bar{u}/v}\, du\, dv = e - \dfrac{1}{e} \approx 2.3504.$

55. (b) Integration by parts:

$$\left[\begin{array}{ll} u = 8\sqrt{3}\,\arcsin\dfrac{1}{\sqrt{3 - x^2}} & dv = dx \\[2mm] du = 8\sqrt{3}\,\dfrac{x}{(3 - x^2)\sqrt{2 - x^2}}\,dx & v = x \end{array}\right]$$

$$A = \left[8x\sqrt{3}\,\arcsin\dfrac{1}{\sqrt{3 - x^2}}\right]_0^1 - 8\sqrt{3}\int_0^1 \dfrac{x^2}{(3 - x^2)\sqrt{2 - x^2}}\,dx$$

$$= 8\sqrt{3}\,\arcsin\dfrac{1}{\sqrt{2}} - 8\sqrt{3}\int_0^1 \dfrac{x^2}{(3 - x^2)\sqrt{2 - x^2}}\,dx$$

$$= 2\pi\sqrt{3} - 8\sqrt{3}\int_0^1 \dfrac{x^2}{(3 - x^2)\sqrt{2 - x^2}}\,dx.$$

Now let $x = \sqrt{2}\sin\theta$:

$$A = 2\pi\sqrt{3} - 8\sqrt{3}\int_0^{\pi/4} \dfrac{2\sin^2\theta}{(3 - 2\sin^2\theta)}\,d\theta$$

$$= 2\pi\sqrt{3} - 8\sqrt{3}\int_0^{\pi/4} \dfrac{1 - \cos 2\theta}{3 - 2\sin^2\theta}\,d\theta = \tfrac{1}{2}\int \dfrac{1 - \cos\phi}{2 + \cos\phi}\,d\phi. \quad (\phi = 2\theta)$$

Finally substitute $u = \tan\left(\dfrac{\theta}{2}\right)$ to (eventually) obtain $A = 4\pi\,(\sqrt{3} - 1).$

Chapter 16: Vector Analysis

Section 16.1

13. $\operatorname{div} \mathbf{F}\,(x,y,z) = 0;$ $\operatorname{curl} \mathbf{F}\,(x,y,z) = \begin{vmatrix} \mathbf{i} & \mathbf{j} & \mathbf{k} \\ \dfrac{\partial}{\partial x} & \dfrac{\partial}{\partial y} & \dfrac{\partial}{\partial z} \\ yz & xz & xy \end{vmatrix} = \langle x - x, y - y, z - z \rangle = \mathbf{0}.$

16. $\operatorname{div} \mathbf{F}\,(x,y,z) = 12;$ $\operatorname{curl} \mathbf{F}\,(x,y,z) = \langle 2,3,1 \rangle.$

22. Write $\mathbf{F} = \langle P, Q, R \rangle$ and $\mathbf{G} = \langle S, T, U \rangle$. Then

$$\begin{aligned} \operatorname{div}(a\mathbf{F} + b\mathbf{G}) &= \operatorname{div} \langle aP + bS, aQ + bT, aR + bU \rangle \\ &= aP_x + bS_x + aQ_y + bT_y + aR_z + bU_z \\ &= a\,(P_x + Q_y + R_z) + b\,(S_x + T_y + U_z) \\ &= (a \operatorname{div} \mathbf{F}) + (b \operatorname{div} \mathbf{G}). \end{aligned}$$

28. Write $\mathbf{F} = \langle P, Q, R \rangle$. Then

$$\operatorname{div}(\operatorname{curl} \mathbf{F}) = \operatorname{div} \langle R_y - Q_z, P_z - R_x, Q_x - P_y \rangle = R_{yx} - Q_{zx} + P_{zy} - R_{xy} + Q_{xz} - P_{yz} = 0$$

provided $R_{yx} = R_{xy}$, $Q_{zx} = Q_{xz}$, and $P_{zy} = P_{yz}$; these equations hold in any region where (say) all second-order partial derivatives are continuous, a condition that normally holds in applications.

34. $\nabla \times \dfrac{1}{r^3}\mathbf{r} = \dfrac{1}{r^3} \begin{vmatrix} \mathbf{i} & \mathbf{j} & \mathbf{k} \\ \dfrac{\partial}{\partial x} & \dfrac{\partial}{\partial y} & \dfrac{\partial}{\partial z} \\ x & y & z \end{vmatrix} = \dfrac{1}{r^3} \langle 0, 0, 0 \rangle = \mathbf{0}.$

37. $\operatorname{div}(r\mathbf{r}) = \operatorname{div}\left(\sqrt{x^2 + y^2 + z^2}\,\langle x, y, z \rangle \right)$

$$= \frac{\partial}{\partial x}\left(x\sqrt{x^2 + y^2 + z^2} \right) + \frac{\partial}{\partial y}\left(y\sqrt{x^2 + y^2 + z^2} \right) + \frac{\partial}{\partial z}\left(z\sqrt{x^2 + y^2 + z^2} \right)$$

$$= 3\sqrt{x^2 + y^2 + z^2} + \frac{x^2}{\sqrt{x^2 + y^2 + z^2}} + \frac{y^2}{\sqrt{x^2 + y^2 + z^2}} + \frac{z^2}{\sqrt{x^2 + y^2 + z^2}} = 3r + r = 4r.$$

40. $\operatorname{grad}\left(r^{10} \right) = \operatorname{grad}\left(\left(x^2 + y^2 + z^2 \right)^5 \right)$

$$= \langle 5\left(x^2 + y^2 + z^2 \right)^4 (2x), 5\left(x^2 + y^2 + z^2 \right)^4 (2y), 5\left(x^2 + y^2 + z^2 \right)^4 (2z) \rangle$$

$$= 10\left(x^2 + y^2 + z^2 \right)^4 \langle x, y, z \rangle = 10r^8 \mathbf{r}.$$

Section 16.2

1. $dx = 4\,dt$, $dy = 3\,dt$, and $ds = 5\,dt$. Also $f\,(t) = (4t - 1)^2 + (3t + 1)^2 = 25t^2 - 2t + 2.$

Therefore $\displaystyle\int_C f\,(x,y)\,ds = \int_{-1}^{1} \left(25t^2 - 2t + 2 \right)(5)\,dt = \frac{310}{3};$

$\displaystyle\int_C f\,(x,y)\,dx = \frac{248}{3}$ and $\displaystyle\int_C f\,(x,y)\,dy = 62.$

4. Note that $ds = dt$. So $\displaystyle\int_C (2x - y)\,ds = \int_0^{\pi/2} (2\sin t - \cos t)\,dt = 1;$

$$\int_C (2x - y)\,dx = \int_0^{\pi/2} \left(2\sin t \cos t - \frac{1 + \cos 2t}{2} \right) dt = \frac{4 - \pi}{4};$$

$$\int_C (2x - y)\,dy = \int_0^{\pi/2} \left(-2\sin^2 t + \sin t \cos t \right) dt = \frac{1 - \pi}{2}.$$

7. Use y as the parameter.

10. If $f(x, y) = \frac{1}{2}x^2 + 2xy - \frac{1}{2}y^2$, then $\nabla f = \langle x + 2y, 2x - y \rangle$. Therefore the given integral is independent of the path. Hence $\int_C (x + 2y)\, dx + (2x - y)\, dy = f(-2, -1) - f(3, 2) = -9$.

13. $\int_C \mathbf{F} \cdot \mathbf{T}\, ds = \int_{t=0}^{\pi} (y\, dx - x\, dy + z\, dz) = \int_0^{\pi} \left(\cos^2 t + \sin^2 t + 4t \right) dt = \pi + 2\pi^2$.

16. C is parallel to $\langle 2, 3, 3 \rangle$, so C may be parametrized by

$$\mathbf{r}(t) = \langle 1, -1, 2 \rangle + \langle 2t, 3t, 3t \rangle = \langle 2t + 1, 3t - 1, 3t + 2 \rangle;$$

that is, we let $x = 2t + 1$, $y = 3t - 1$, and $z = 3t + 2$ for $0 \le x \le 1$. Then $ds = \sqrt{22}\, dt$, so

$$\int_C xyz\, ds = \int_0^1 (2t + 1)(3t - 1)(3t + 2)\sqrt{22}\, dt = 7\sqrt{22}.$$

19. If the wire has constant density k, then the total mass of the wire is $M = \pi a k$ and the mass of a segment of length ds is $dm = k\, ds$. Parametrize the wire as follows:

$$x = a \cos \theta, \quad y = a \sin \theta, \quad 0 \le \theta \le \pi.$$

Then compute $M_x = \int_C y\, dm$.

22. $I_z = \int_C (x^2 + y^2)\, dm = 90 k \pi^2 = 9M$.

25. (a) $W = \int_0^1 \frac{ky}{1 + y^2}\, dy$; (b) $W = \int_1^0 \frac{kx}{1 + x^2}\, dx$.

28. We assume that the particle moves uniformly, so that its path may be parametrized as follows: $\mathbf{r}(t) = \langle \cos t, \sin t \rangle$, $0 \le t \le 2\pi$. Then

$$W = \int_0^{2\pi} \langle -\sin t, \cos t \rangle \cdot \langle -\sin t, \cos t \rangle\, dt = \int_0^{2\pi} \left(\sin^2 t + \cos^2 t \right) dt = 2\pi.$$

31. Parametrization: $x = 100 \sin t$, $y = 100 \cos t$, $\pi/2 \ge t \ge 0$.

$$W = \int_{\pi/2}^0 \langle 0, -150 \rangle \cdot \langle 100 \cos t, -100 \sin t \rangle\, dt = \left[-15{,}000 \cos t \right]_0^{\pi/2} = 15{,}000 \quad \text{(ft-lb)}.$$

Section 16.3

1. If $\nabla f = \mathbf{F} = \langle 2x + 3y, 3x + 2y \rangle$ then $f_x(x, y) = 2x + 3y$, so $f(x, y) = x^2 + 3xy + g(y)$. Then $3x + g'(y) = f_y(x, y) = 3x + 2y$, so we may choose $g(y) = y^2$. Answer: One potential function for $\mathbf{F}$ is $f(x, y) = x^2 + 3xy + y^2$.

4. $f(x, y) = x^2 y^2 + x^3 + y^4$

10. $f(x, y) = e^x \sin y + x \tan y$

22. $\left[xe^y + ye^x \right]_{(0,0)}^{(1,-1)} = \frac{1}{e} - e$

28. Choose g such that $\mathbf{F} = \nabla g$. Then $P = g_x$, $Q = g_y$, and so on. Assume as usual that the second-order partial derivatives of g are continuous to obtain the desired conclusion.

Section 16.4

1. $\int_{-1}^1 \int_{-1}^1 (2x - 2y)\, dy\, dx = \int_{-1}^1 4x\, dx = 0$.

4. $\int_0^1 \int_{x^2}^{x} (y + 2y)\, dy\, dx = \frac{1}{5}.$

7. $\int_0^{\pi} \int_0^{\sin x} 1\, dy\, dx$

10. $Q_x - P_y = 0$, so the value of the integral is zero.

13. Let C denote the circle with the parametrization given in the problem. The area of the circular disk it bounds is given by

$$A = \oint_C x\, dy = \oint_C (a \cos t)(a \cos t)\, dt = a^2 \int_0^{2\pi} \tfrac{1}{2}(1 + \cos 2t)\, dt = \pi a^2.$$

16. The area is $\frac{5}{12}$.

22. (a) $\left(0, \dfrac{4a}{3\pi}\right);$ (b) $\left(\dfrac{4a}{3\pi}, \dfrac{4a}{3\pi}\right).$

Section 16.5

1. Parametrization: $u = x$, $v = y$. Use $\mathbf{r}(u, v) = \langle u, v, 1 - u - v\rangle$. Then $|\mathbf{r}_u \times \mathbf{r}_v| = \sqrt{3}$. Let D denote the first-quadrant region bounded by the x-axis, the y-axis, and the line $x + y = 1$. Then

$$\iint_S f(x, y, z)\, dS = \iint_D xy(1 - x - y)\sqrt{3}\, dx\, dy = \sqrt{3}\int_0^1 \int_0^{1-x} (xy - x^2 y - xy^2)\, dy\, dx = \tfrac{1}{120}\sqrt{3}.$$

4. In spherical form the function f has the formula $f(\rho, \phi, \theta) = \rho^3 \sin^2 \phi \cos \phi$. A suitable parametrization for S is

$$\mathbf{r}(\phi, \theta) = \langle \sin \phi \cos \theta, \sin \phi \sin \theta, \cos \phi\rangle, \quad 0 \le \theta \le 2\pi, \quad 0 \le \phi \le \pi/2.$$

Next you should find that

$$\mathbf{r}_\phi \times \mathbf{r}_\theta = \langle \sin^2 \phi \cos \theta, \sin^2 \phi \sin \theta, \cos \phi \sin \phi\rangle;$$

it follows that $dS = |\sin \phi|\, d\phi\, d\theta = \sin \phi\, d\phi\, d\theta$. Thus the surface integral becomes

$$\int_0^{2\pi} \int_0^{\pi/2} \rho^3 \sin^3 \phi \cos \phi\, d\phi\, d\theta.$$

With the aid of the fact that $\rho = 1$, we find the value of this integral to be $\pi/2$.

Alternative solution: Describe S as follows: $z = h(x, y) = \sqrt{1 - x^2 - y^2}$ for $x^2 + y^2 \le 1$. Show that $dS = \dfrac{1}{z}\, dx\, dy$. Then show that the integral takes the form

$$\iint_C (x^2 + y^2)\, dx\, dy$$

where D is the disk $x^2 + y^2 \le 1$. Use polar coordinates.

7. Parametrize S by $\mathbf{r}(u, v) = \langle u, v, 3u + 2\rangle$ on the disk D: $u^2 + v^2 \le 4$. Then

$$\frac{\partial(y, z)}{\partial(u, v)} = -3, \quad \frac{\partial(z, x)}{\partial(u, v)} = 0, \quad \text{and} \quad \frac{\partial(x, y)}{\partial(u, v)} = 1.$$

Therefore

$$\iint_S \mathbf{F} \cdot \mathbf{n}\, dS = \iint_D (-6y + 3z)\, du\, dv = \iint_D (-6v + 9u + 6)\, du\, dv$$

$$= \int_0^{2\pi} \int_0^2 (6 - 6r \sin \theta + 9r \cos \theta)\, r\, dr\, d\theta = \int_0^{2\pi} (12 - 16 \sin \theta + 24 \cos \theta)\, d\theta = 24\pi.$$

10. 44π

13. Checkpoints: $M = \frac{1}{6}\pi\delta\left((4a^2+1)^{3/2}-1\right)$; $M_{xy} = \frac{1}{60}\pi\delta\left(1+(24a^4+2a^2-1)\sqrt{4a^2+1}\right)$.

16. $I_z = a\delta\displaystyle\int_0^{2\pi}\int_0^{2\pi}(b+a\cos\psi)^3\,d\psi\,d\theta$.

19. Checkpoints: The flux across the lower surface is -243π and the flux across the upper surface is 1701π.

Section 16.6

1. First, $\operatorname{div}\mathbf{F} = 3$. So $\displaystyle\iiint_B \operatorname{div}\mathbf{F}\,dV = \iiint_B 3\,dV = (3)\left(\frac{4}{3}\right)(\pi)(1)^3 = 4\pi$.

   A unit normal is $\mathbf{n} = \langle x,y,z\rangle$, and $\mathbf{F}\cdot\mathbf{n} = 1$. So $\displaystyle\iint_S \mathbf{F}\cdot\mathbf{n}\,dS = (1)(4\pi)(1)^2 = 4\pi$.

   Both values are 4π; this verifies the divergence theorem in this special case.

4. Here, $\operatorname{div}\mathbf{F} = y+z+x$; on the face where $z = 0$, $0 \le x \le 2$, and $0 \le y \le 2$, we have $\mathbf{F}\cdot\mathbf{n} = 0$. On the opposite face we find that $\mathbf{F}\cdot\mathbf{n} = 2x$. The surface integral of $\mathbf{F}\cdot\mathbf{n}$ on that face is equal to 8. Similar results hold on the other four faces, and both integrals in the divergence theorem are equal to 24.

7. Let C denote the solid cylinder. Then
$$\iint_S \mathbf{F}\cdot\mathbf{n}\,dS = \iiint_C (3x^2+3y^2+3z^2)\,dV = 3\int_0^{2\pi}\int_0^3\int_{-1}^4 (r^2+z^2)\,r\,dz\,dr\,d\theta.$$

10. Let B denote the solid bounded by S; note that $\operatorname{div}\mathbf{F} = x^2+y^2$. So
$$\iint_S \mathbf{F}\cdot\mathbf{n}\,dS = \iiint_B \operatorname{div}\mathbf{F}\,dV = \int_0^{2\pi}\int_0^3\int_{r^2}^9 r^3\,dz\,dr\,d\theta = \frac{243}{2}\pi.$$

Section 16.7

1. Because $\mathbf{n}$ is to be the upper unit normal, we have
$$\mathbf{n} = \mathbf{n}(x,y,z) = \tfrac{1}{2}\langle x,y,z\rangle.$$

   The boundary curve C of the hemispherical surface S has the parametrization
$$x = 2\cos\theta,\quad y = 2\sin\theta,\quad 0 \le \theta \le 2\pi.$$

   Therefore
$$\iint_S (\operatorname{curl}\mathbf{F})\cdot\mathbf{n}\,dS = \int_C 3y\,dx - 2x\,dy + xyz\,dz = \int_0^{2\pi}(-12\sin^2\theta - 8\cos^2\theta)\,d\theta = -20\pi.$$

4. Parametrize the boundary curves as follows:
$$C_1:\quad x = \cos t,\quad y = \sin t,\quad z = 1,\quad 0 \le t \le 2\pi;$$
$$C_2:\quad x = \cos t, y = -\sin t,\quad z = 3,\quad 0 \le t \le 2\pi.$$

   Then
$$\iint_S (\operatorname{curl}\mathbf{F})\cdot\mathbf{n}\,dS = \int_{C_1}\mathbf{F}\cdot\mathbf{T}\,ds + \int_{C_2}\mathbf{F}\cdot\mathbf{T}\,ds = \int_0^{2\pi}(2\sin^2 t - 2\cos^2 t)\,dt = 0.$$

7. Parametrize S (the elliptical region bounded by C) as follows:
$$x = z = r\cos t, \quad y = r\sin t, \quad 0 \le r \le 2, \quad 0 \le t \le 2\pi.$$
Then $\mathbf{r}_r \times \mathbf{r}_t = \langle -r, 0, r \rangle$, $dS = r\sqrt{2}\, dr\, dt$; the upper unit normal for S is
$$\mathbf{n} = \tfrac{1}{2}\sqrt{2}\,\langle -1, 0, 1 \rangle,$$
and $\operatorname{curl} \mathbf{F} = \langle 3, 2, 1 \rangle$. So $(\operatorname{curl} \mathbf{F}) \cdot \mathbf{n} = -\sqrt{2}$. Hence
$$\int_C \mathbf{F} \cdot \mathbf{T}\, ds = \iint_S (\operatorname{curl} \mathbf{F}) \cdot \mathbf{n}\, dS = \int_0^{2\pi} \int_0^2 (-2r)\, dr\, dt = -8\pi.$$

10. Let E be the ellipse bounded by C. Now E lies in the plane with equation $-y + z = 0$, so has upper unit normal
$$\mathbf{n} = \tfrac{1}{2}\sqrt{2}\,\langle 0, -1, 1 \rangle.$$
Next, $\operatorname{curl} \mathbf{F} = \langle -2z, -2x, -2y \rangle$, so $(\operatorname{curl} \mathbf{F}) \cdot \mathbf{n} = \sqrt{2}\,(x - y)$. The projection of E into the xy-plane may be described this way:
$$x^2 + (y - 1)^2 = 1, \quad z = 0;$$
alternatively, by $r = 2\sin t$ $(0 \le t \le \pi)$, $z = 0$. Thus
$$\int_C \mathbf{F} \cdot \mathbf{T}\, ds = \iint_E (\operatorname{curl} \mathbf{F}) \cdot \mathbf{n}\, dS = \iint_E \sqrt{2}\,(x - y)\, dS.$$
A parametrization of E is
$$\mathbf{w}\,(r, t) = \langle r\cos t, r\sin t, r\sin t \rangle, \quad 0 \le t \le \pi, \quad 0 \le r \le 2\sin t.$$
Next we find that $\mathbf{n} = \mathbf{w}_r \times \mathbf{w}_t = \langle 0, -r, r \rangle$. Therefore $dS = r\sqrt{2}\, dr\, dt$. Consequently,
$$\iint_E \sqrt{2}\,(x - y)\, dS = \int_0^\pi \int_0^{2\sin t} \sqrt{2}\, r\,(\cos t - \sin t)\, r\sqrt{2}\, dr\, dt.$$
Answer: -2π.

22. Another of the glories of Western civilization.
$$L = \iint_S (\mathbf{r} - \mathbf{r}_0) \times (-\rho g z \mathbf{n})\, dS = \rho g \iiint_V \operatorname{curl}(z\,(\mathbf{r} - \mathbf{r}_0))\, dV \quad \text{(by Problem 20)}.$$
But $\operatorname{curl}(z\,(\mathbf{r} - \mathbf{r}_0)) = (\nabla z) \times (\mathbf{r} - \mathbf{r}_0) + z\,(\operatorname{curl}(\mathbf{r} - \mathbf{r}_0))$. It follows immediately that $\nabla z = \mathbf{k}$ and that $\operatorname{curl}(\mathbf{r} - \mathbf{r}_0) = \mathbf{0}$. Thus
$$\mathbf{L} = \rho g \iiint_V \mathbf{k} \times (\mathbf{r} - \mathbf{r}_0)\, dV = \rho g \mathbf{k} \times \left(\iiint_V \mathbf{r}\, dV - \mathbf{r}_0 V \right).$$
consequently $\mathbf{L} = \mathbf{0}$ as desired, because $\mathbf{r}_0 = \dfrac{1}{V} \iiint_V \mathbf{r}\, dV$.

Chapter 16 Miscellaneous

1. C is part of the graph of $y = \tfrac{4}{3}x$ and $ds = \tfrac{5}{3}\, dx$. So $\displaystyle\int_C (x^2 + y^2)\, ds = \int_0^3 \left(x^2 + \left(\tfrac{4}{3}x\right)^2 \right) \tfrac{5}{3}\, dx = \tfrac{125}{3}$.

4. 115

7. If $\nabla\phi = \langle x^2 y, xy^2 \rangle$ then $\phi_x = x^2 y$, so that $\phi\,(x, y) = \tfrac{1}{3}x^3 y + g\,(y)$. Thus $\phi_y = \tfrac{1}{3}x^3 + g'\,(y) = xy^2$. This is impossible unless x is constant, but x is not constant on C.

10. Parametrize C: $x = t$, $y = t^2$, $z = t^3$, $1 \le z \le 2$.
$$W = \int_C P\, dx + Q\, dy + R\, dz = \int_C z\, dx - x\, dy + y\, dz = \int_1^2 (t^3 - 2t^2 + 3t^4)\, dt = \tfrac{1061}{60}.$$

16. Note that if $P_y < Q_x$ at some point of D, then $Q_x - P_y > 0$ on a small region R surrounding that point, so that
$$\iint_R (Q_x - P_y)\, dA > 0.$$

19. Checkpoints: The flux across the upper surface is 60π and the flux across the lower surface is 12π.

Appendices
Appendix A

1. $40 \cdot \dfrac{\pi}{180} = \dfrac{2\pi}{9}$ (rad)

4. $210 \cdot \dfrac{\pi}{180} = \dfrac{7\pi}{6}$

7. $\dfrac{2\pi}{5} \cdot \dfrac{180}{\pi} = 72°$

10. $\dfrac{23\pi}{60} \cdot \dfrac{180}{\pi} = 69°$

x	$\sin x$	$\cos x$	$\tan x$	$\cot x$	$\sec x$	$\csc x$
13.	$\frac{7}{6}\pi$	$-\frac{1}{2}$	$-\frac{1}{2}\sqrt{3}$	$\frac{1}{3}\sqrt{3}$	$\sqrt{3}$	$-\frac{2}{3}\sqrt{3}$	-2

16. $\sin x = 1$ for $x = \frac{1}{2}(4n+1)\pi$ where n is an integer.

19. $\cos x = 1$ when x is an integral multiple of 2π.

22. $\tan x = 1$ for $x = \frac{1}{4}(4k+1)\pi$ where k is an integer.

25. See the figure at the right.

$\sin x = -\frac{3}{5}, \cos x = \frac{4}{5}, \tan x = -\frac{3}{4},$

$\csc x = -\frac{5}{3}, \sec x = \frac{5}{4}, \cot x = -\frac{4}{3}.$

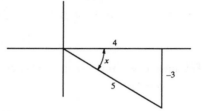

31. $\sin(11\pi/6) = \sin(-\pi/6) = -\sin(\pi/6) = -1/2$

34. $\cos(4\pi/3) = -\cos(\pi/3) = -1/2$

37. (a) $\cos\left(\dfrac{\pi}{2} - \theta\right) = \cos\left(\dfrac{\pi}{2}\right)\cos\theta + \sin\left(\dfrac{\pi}{2}\right)\sin\theta = 0 \cdot \cos\theta + 1 \cdot \sin\theta = \sin\theta.$

40. $\tan(\pi \pm \theta) = \tan(\pm\theta)$ (because the tangent function is periodic with period π) and $\tan(\pm\theta) = \pm\tan(\theta)$ because the tangent function is odd (the quotient of an odd function and an even function).

43. $3\sin^2 x - 1 + \sin^2 x = 2:$
$4\sin^2 x = 3;$
$\sin x = \pm\frac{1}{2}\sqrt{3}$
$x = \frac{1}{3}\pi, x = \frac{2}{3}\pi.$

46. $2\sin^2 x + \cos x = 2:$
$2 - 2\cos^2 x + \cos x = 2;$
$(\cos x)(2\cos x - 1) = 0;$
$\cos x = 0$ or $\cos x = \frac{1}{2};$
$x = \frac{1}{2}\pi, x = \frac{1}{3}\pi.$

Appendix B

1. Given $\epsilon > 0$, choose $\delta = \epsilon$. If $0 < |x - a| < \delta$ then $|x - a| < \epsilon$. Therefore $\lim\limits_{x \to a} x = a$.

4. Given $\epsilon > 0$, choose $\delta = \epsilon/2$. If $0 < |x - (-3)| < \delta$ then $|2x + 6| < 2\delta$, so $|(2x + 1) - (-5)| < \epsilon$. Therefore, by definition, $\lim\limits_{x \to -3}(2x + 1) = -5$.

7. Given $\epsilon > 0$, let $\delta = \epsilon/k$ where you will choose a suitably large positive integer later in the proof so that the inequalities will work. Then go back through the proof and replace k by this integer. Write it neatly and turn it in for a perfect grade. This is a standard technique in ϵ-δ proofs.

10. The desired inequality $\left|\dfrac{1}{\sqrt{x}} - \dfrac{1}{\sqrt{a}}\right| < \epsilon$ follows from $|\sqrt{a} - \sqrt{x}| < \epsilon\sqrt{ax}$, and we keep x from becoming negative or zero (or too close to zero) by requiring that $a/2 < x < 3a/2$. We obtain the inequality $|\sqrt{a} - \sqrt{x}| < \epsilon\sqrt{ax}$ from

$$|\sqrt{a} + \sqrt{x}| \, |\sqrt{a} - \sqrt{x}| < \epsilon\sqrt{ax} \left(\sqrt{a} + \sqrt{x}\right);$$

that is,

$$|x - a| < \epsilon\sqrt{ax} \left(\sqrt{a} + \sqrt{x}\right).$$

It suffices that

$$\delta < \epsilon\sqrt{ax} \left(\sqrt{a} + \sqrt{x}\right),$$

and this can be assured by choosing $\delta < a\epsilon/2$, as this implies the last displayed inequality. So choose δ to be the minimum of $a/2$ and $a\epsilon/2$.

13. First deal with the case $L > 0$. The desired inequality follows from $|f(x) - L| < \epsilon L f(x)$. But $f(x)$ can be forced to lie between $L/2$ and $3L/2$ by choosing δ sufficiently small, and now the choice of δ is clear: Merely ensure that $\delta < L^2\epsilon/2$.

16. Let $\epsilon = \frac{1}{2}f(a)$. Note that $f(x) \to f(a)$ as $x \to a$ by hypothesis. Now choose δ such that

$$|f(x) - f(a)| < \epsilon \quad \text{if} \quad 0 < |x - a| < \delta.$$

For such x, we have $-\epsilon < f(x) - 2\epsilon < \epsilon$, so $\epsilon < f(x) < 3\epsilon$. Therefore $f(x) > 0$ if x is in the interval $(a - \delta, a + \delta)$.

Appendix F

1. $\displaystyle\int_0^1 x^2 \, dx = \frac{1}{3}.$

4. $\displaystyle\int_0^3 \frac{x}{\sqrt{16 + x^2}} \, dx = \left[\sqrt{16 + x^2}\right]_0^3 = 1.$

7. $\displaystyle\int_0^2 \sqrt{x^4 + x^7} \, dx = \left[\frac{2}{9}\left(1 + x^3\right)^{3/2}\right]_0^2 = \frac{52}{9}.$

Software: *Derive* 2.56, *Mathematica* 2.2, WordPerfect 5.1, Adobe Illustrator 3.2, and (most important) TeX 3.0.

Hardware: Zenith Z-386SX, Macintosh IIfx, Apple Laserwriter IIf, La Cie Pocket-85, and the North Georgia Mountains.